Science and Fiction

For further volumes:

http://www.springer.com/series/11657

Science and Fiction – A Springer Series

This collection of entertaining and thought-provoking books will appeal equally to science buffs, scientists and science-fiction fans. It was born out of the recognition that scientific discovery and the creation of plausible fictional scenarios are often two sides of the same coin. Each relies on an understanding of the way the world works, coupled with the imaginative ability to invent new or alternative explanations—and even other worlds. Authored by practicing scientists as well as writers of hard science fiction, these books explore and exploit the borderlands between accepted science and its fictional counterpart. Uncovering mutual influences, promoting fruitful interaction, narrating and analyzing fictional scenarios, together they serve as a reaction vessel for inspired new ideas in science, technology, and beyond.

Whether fiction, fact, or forever undecidable: the Springer Series "Science and Fiction" intends to go where no one has gone before!

Its largely non-technical books take several different approaches. Journey with their authors as they

- Indulge in science speculation – describing intriguing, plausible yet unproven ideas;
- Exploit science fiction for educational purposes and as a means of promoting critical thinking;
- Explore the interplay of science and science fiction – throughout the history of the genre and looking ahead;
- Delve into related topics including, but not limited to: science as a creative process, the limits of science, interplay of literature and knowledge;
- Tell fictional short stories built around well-defined scientific ideas, with a supplement summarizing the science underlying the plot.

Readers can look forward to a broad range of topics, as intriguing as they are important. Here just a few by way of illustration:

- Time travel, superluminal travel, wormholes, teleportation
- Extraterrestrial intelligence and alien civilizations
- Artificial intelligence, planetary brains, the universe as a computer, simulated worlds
- Non-anthropocentric viewpoints
- Synthetic biology, genetic engineering, developing nanotechnologies
- Eco/infrastructure/meteorite-impact disaster scenarios
- Future scenarios, transhumanism, posthumanism, intelligence explosion
- Virtual worlds, cyberspace dramas
- Consciousness and mind manipulation

Wallace Kaufman • David Deamer

The Hunt for FOXP5

A Genomic Mystery Novel

Wallace Kaufman
Newport, Oregon
USA

David Deamer
Santa Cruz, California
USA

ISSN 2197-1188
Science and Fiction
ISBN 978-3-319-28960-1
DOI 10.1007/978-3-319-28961-8

ISSN 2197-1196   (electronic)

ISBN 978-3-319-28961-8   (eBook)

Library of Congress Control Number: 2016935426

Cover illustration: 3d illustration of viruses attacking nerve cells, concept for Neurologic Diseases, tumors and brain surgery. By Ralwel/Shutterstock.com.

This Springer imprint is published by Springer Nature
The registered company is Springer International Publishing AG Switzerland

For Meruert Yegemberdieva, a model of so much that is rich and wonderful in Kazakhstan and in our heroine Avalon, and with thanks for their welcome and friendship to her parents Zholdas and Rita, her grandparents Gulya and Zautbek.

Contents

Part I

The Novel

The Hunt for FOXP5

*Like Sisyphus rolling his boulder up to the top of the hill
only to have it tumble down again, the human gene pool
creates hereditary genius in many ways in many places only
to have it come apart in the next generation.*

E. O. Wilson 1978

Chapter 1

December 2020

Students filled every seat in the University of Oregon's biology lecture hall. Latecomers were sitting on steps, and a few students in T-shirts and jeans were standing behind the last row. The usual low murmur of conversation died away when the lights dimmed and a strange image appeared on the screen above the stage. Enthusiastic clapping erupted as the speaker emerged from behind the curtain and strode confidently to the podium. She stepped onto a 6 in. wooden box put there for her, and gripped both sides of the podium as if to steady it. She stood quietly, smiling at the audience until the applause quieted. Then she released the podium and directed a laser toward the images behind her.

"You see four faces on the screen. The first is obviously a baboon. The second you might know is a bonobo, but the third you will have to guess—it's Homo erectus, one of our primate ancestors." The speaker then used the laser beam to circle the fourth image, an illustration from *National Geographic* showing a nude female holding an infant. She was obviously human, but had dark skin and fine fur on her arms and legs. "This is your mother, and mine. We'll call her Ma."

A twitter of laughter rose in a wave of sound that crested and gradually spent itself in the darkened auditorium. The faces faded from the screen and a spotlight circled the small, trim figure in a white pants suit, her face framed by a neat helmet of black hair. Michelle Murphy at 36 did not stand out in a

© Springer International Publishing Switzerland 2016
W. Kaufman, D. Deamer, *The Hunt for FOXP5*, Science and Fiction,
DOI 10.1007/978-3-319-28961-8_1

crowd or a classroom. She looked younger than her age—large dark eyes beneath dark eyebrows and black hair, sturdy but not stout. Her anthropologist colleague, Don Koskin, once suggested that her hair color and eyes reflected genes from ancient Ierne, the land Romans called Hibernia, today's Ireland. Michelle, who knew even more about genes, told him he was right; her father had the R1b-S-4 markers, the so-called Irish clade. She was given to wearing dark clothes or light clothes but not bright clothes. Overall she projected to students and colleagues a person both private and confident.

Referring to the slide, Professor Murphy said, "I have several reasons for calling her Ma, and I will explain them. First, I must tell you how we found your mother and mine. I owe this to a young woman who is in the audience today—Julie Flanagan. When I met Julie four years ago on a primate fossil dig in northern Ethiopia, the crew called her The Cooler."

"The Cooler", sitting in the first row grinned with pride and affection. Beside her, holding The Cooler's hand lightly, sat a skinny, nut brown 12 year old with long jet black hair. Michelle smiled at her daughter, Avalon, who was a pleasure in her heart, a reminder of her missing husband, and a constant puzzle for her brain.

During the laughter following the first slide, Michelle had scanned the audience. Behind her daughter and Julie she saw students from her classes, and several of her faculty colleagues. She also noticed two oddly dressed people she didn't recognize. Oddly dressed because students don't wear suits, and most professors don't wear suits. The handsome, tall, athletic young man in the dark suit and tie was too old to be a student, too young to be faculty. Next to him sat a short, sturdy woman who probably once had the build of a chunky soccer player. Now she was a solid matron in a dark purple blazer, a Roman nose. They had smiled, but they had not laughed. Michelle hoped they were not selling lab equipment or textbooks or insurance. Only after the lecture would Michelle find out that they had come to see her on business that would force her to learn more than she ever wanted to know about biological warfare. What she knew at the moment was that she had just started to lecture about how the world had changed some 160,000 years ago, but her story began just 2 years before in a hot Ethiopian desert.

Chapter 2

August 2016

Julie "The Cooler" Flanagan rarely got excited, so when she shouted—"Got an arm, an arm bone!" the other diggers kneeling in their assigned squares of dried riverbed sediments stopped work and looked to her. The dig was in an area of Ethiopia called the Middle Awash Valley where the blazing eye of summer sun sucked life out of the land, and those who dug in it became covered with its ancient dust until they looked as if they had emerged from the stuff they were digging into. They had not begun calling her The Cooler because of a calm demeanor, however. Her intensely freckled face looked excited even when she was asleep. Cooler had come from "The Arizona Cooler." She insisted, sometimes tediously, that the best way to stay cool in the searing African sun was to wear a thick cotton hat and keep it soaked with water. That was her version of the well known Arizona cooler—a field rigged solar refrigerator that cooled its contents by evaporation. That classical cooler was a box draped in many layers of burlap topped by a large can with tiny holes in the bottom. The water in the can slowly leaked onto the cloth, keeping it wet so that as the sun evaporated the water, it kept the drinks and sandwiches inside cool. Julie's cerebral cooler occasioned daily jokes—"Where's the water can?" or "No wonder she never gets excited." Or "It would work better if you cut off that blonde insulation underneath." The two senior undergraduate men who were privileged to be on the dig thought the tall, trim, young blonde was just plain cool.

The small group at the site was led by Dr. Arthur Adamski. During the single month allowed by the government for the dig, ten students worked for the entire time, two couples, Julie, and five single men. The single men and couples were all graduate students handpicked by Adamski from many volunteers who applied to join the dig. Julie was a freshman scholarship student at City University of New York, recommended so enthusiastically by her award winning high school social studies teacher that Adamski could not refuse. Besides, that teacher, George Pappas, had often cleaned Adamski's pockets when they had played poker in grad school.

Occasional visitors came up onto the plateau trailing a rooster tail of dust behind a hired SUV. A few days earlier Michelle Murphy, a colleague of Adamski's from her time at the University of Indiana, had come to visit with her 8 year old daughter Avalon. Adamski had served as thesis supervisor for Michelle's late husband, and the three had become friends. The daughter of the Irish looking, compactly built Dr. Murphy was tall, gangling, agile, and intensely curious. Her skin was the color of cherry wood. A keen student of physical anthropology would easily identify this brown eyed daughter as Central Asian. If she were Dr. Murphy's natural daughter, she was all father and no mother, but she was so entirely Asian, she must be adopted. No one asked. They were pleased that Avalon never seemed bored, could hunker for an hour watching a painstaking excavation, and they were amused that after work the girl would pick up almost any book or magazine, even a scholarly journal, and lose herself in it. One of the grad students asked her the first night as they sat around the table after supper, "Do you understand that stuff?"

Avalon had replied, "Sure, don't you?"

Everyone laughed except Avalon and the student, who looked a bit miffed.

Michelle laughed too but thought, *She understands the book, but she may not understand the people.*

Julie was the youngest student on the dig, but 2 years earlier as a Rotary Exchange student in South Africa she had volunteered on an excavation and been credited with finding the first fragments of human-like bones—a few teeth, a couple of toes, enough of a hip socket to model the complete cup and the ball that would have fit into it. In Ethiopia her fellow diggers quickly learned she had "the eye." She countered by saying, "It's in the brain. That's where you learn to see the difference between matrix and fossil. You form a mental template." She did it easily. Others took all season and some never got it. Her eye had established this dried up river bed as potentially rich in revelations about the origins of humanity. When she had shouted *arm bone*, someone had called back, "Cool, Cooler."

Julie called across the dry wash to the tents where "Dr. Art" the dig supervisor worked with his lenses and microscopes and cameras. Dr. Arthur Adamski stooped as he emerged from under the tent flap and straightened his shirtless six foot six frame just outside the shelter, shielding his eyes from the brilliant sunlight as he searched for Julie. His right side bore a scatter of purple scars where he had been hit by shrapnel in Vietnam. His full head of white hair shone in the sun. As he crossed the rocky ground, he looked like a large spider monkey with his hairy chest and back. Diggers sometimes called the fossil people "Adamski's little children." Little because those early African hominids stood barely 4 ft tall and Adamski was well over 6 ft.

He hunkered down next to Julie who was preparing her find for his examination with rapid but careful strokes of a paint brush, flicking away as much of the crumbly matrix as possible to make her prize clean and obvious. She looked up at Dr. Art. "It's a hominid radius isn't it?"

He studied the exposed end closely without saying anything. By this time other diggers had gathered around, waiting Dr. Art's pronouncement on Cooler's find. Michelle and Avalon also watched. Dr. Murphy smiled down at Julie. Julie and Michelle were the only single women on the site, both serious about their science, and they had quickly become friends. Dr. Art stared intently at the protruding bone. "I wish it were, but something says it's not. Too small, unless it was a child." He picked up the triangular mason's trowel that lay among her tools and carefully picked at the matrix around the bone. Now and then he blew away more of the dust. It stuck in the sweat on his face and turned him a light tan.

While Dr. Art was working, Julie leaned back, surveying the local area, looking for context. Her intense eyes stared out from her own brown mask of dust as if she had acquired some of the antiquity of the place. Then she noticed Avalon standing a few feet away, looking at her and silently pointing down at what looked like just more hardened desert sand. Julie followed her eyes and realized she might have missed something, mostly buried but showing a bit of ivory through the tan sediment. She nodded at the girl just as Adamski finished clearing the loose matrix around her first find. In the mudstone Adamski had cleaned, she could see the outline of a curved canine tooth far too large to be human, lodged next to the arm bone. "I'd say what you have are pieces of a baboon. Sorry Julie, but go ahead, get it out and we'll take a good look."

Dr. Art stood up, and the knot of diggers turned to go to their squares. He called after them, "Take a 15 minute water and rest break while you're up."

Michelle had been using her iPhone to capture images of Julie's find, and quickly snapped a photo of Dr. Art's face as he spoke, outlined by sunlight against the cloudless Ethiopian sky. She knew it would be a keeper, to be added to a digital album she called "Modes and Moods of Humankind." She and Avalon stayed behind, and they both sat down next to Julie, sharing sips of water from her canteen.

Julie stared at the bone and said, "Professor Murphy, what do you think? Is this just a baboon?"

"Julie, I'm Michelle, remember? Just Michelle."

Julie looked up, smiled gratefully and nodded.

Michelle said, "I've known Art since my husband was his graduate student. And he's very careful with words. I'm sure you noticed how he uses qualifiers—'maybe, if, assuming, I think'—all in the conditional tense. At Indiana everyone used to call him Dr. Conditional. Anyway, go ahead and see if you can get the whole thing out. Maybe he'll change his mind." With that, Michelle patted Julie's shoulder, then stood and took her daughter's hand. "Now let's go find our own bone."

As Avalon and her mother returned to their assigned area, the girl looked back at Julie, who winked and used her thumb to point over her shoulder at the bit of white the girl had discovered, signaling that she would give it her full attention soon. She was rewarded with a big smile. Julie wondered what was behind those beautiful dark eyes.

Julie understood the patience needed for this kind of treasure hunting, the patience to continue through hours and days of nothing but heat and dust, rock and gravel and dirt, the patience to keep going when hope turns to disappointment. She pitied those who never seemed to develop the mental templates that distinguished biology from geology—bone from stone, stone tools from naturally broken stones. Yet the girl, Avalon, had developed the ability almost overnight. And that despite the fact that she seemed to have a lazy wandering left eye. Julie hoped it could be corrected.

Today, fortunately, was full of treasure. Within half an hour Julie had the arm bone completely exposed on a small plateau of earth, and the mapper had noted its position and depth in the dig. She held her plastic bottle of Vinac B-15 preservative upside down above the fossil and slowly squeezed the liquid drop by drop through the pointed nose onto the bone where it sank in and began the hardening process that would keep the prize from crumbling. With plaster and burlap strips she made a frame around the oblong plateau, then completed her field note form. While the plaster was hardening, Julie crawled over to the area where Avalon had been pointing down and began flicking away the matrix.

Over the next hour a broken and mostly flattened skull emerged, with the upper and lower jawbones largely intact. It was clearly a baboon skull. And there she saw the treasure. Clenched between the jaws was another arm bone, this one quite unlike the first, at least to the trained eye, and she "had the eye." This time before calling for Dr. Art or making a triumphant noise, she set to work like a careful sculptor until the skull with its mouthful was sitting clean and clear on top of the ground beneath it. When she had finished she looked up and was startled to see Avalon hunkered to one side watching intently.

As coolly as possible she called down to the tent, "Dr. Art? Come have a look, please." The other diggers glanced up, but Julie had positioned herself and her tool bag to block their view. Dr. Art was slower to emerge this time, and when he did, Julie saw that he had taken time to wash dust off his face and out of his eyes. A moment later he again kneeled by her side. After a few seconds, very quietly he said, "Bingo, Cooler. Hang on just a minute." He called over the Tanzanian mapper and instructed him how to diagram Julie's find, then went to the tent for his camera and tripod. When he returned, he knelt next to Julie and took a dozen shots from various angles and magnifications. Finally he stood up, pulled a whistle from his pocket and blew it. When he had attention across the dig site at all the sampling squares, he shouted, "Come take a look."

Everyone knew it must be one of Adamski's children—well, a piece of one. Adamski's Children appeared only in pieces. To the most intense students, like Cooler, the pieces of people conjured the faintest sense of ghosts hovering in the dust. Julie thought of them as her most distant relatives. The group gathered around and stared silently down at a remnant of pre-history that hinted at a dramatic confrontation between humans and baboons perhaps 200,000 years ago. Michelle caught Julie's eye, smiled, nodded, and with her right fist, hit an invisible nail a sound blow. Both of them knew that this once-in-a-lifetime moment would be associated with Julie's name for years to come. It would put her in text books and on web sites.

After everyone had had a look and had congratulated Cooler and complimented her on having "the eye," Julie stood up and said, "Everybody, Dr. Art has told us many times that anyone who makes a significant contribution to a discovery should get credit for it, and I want you all to know that there is someone here who must be recognized." Julie parted the little group and stood behind Michelle and Avalon placing both hands on Avalon's shoulders. "Please congratulate Avalon Murphy."

Everyone applauded politely, a kind of patronize-the-kid applause. Julie went on, "While Dr. Art and I were messing around with my first baboon bone, Avalon spotted a tiny bit of white a few feet away. She didn't say a thing, just pointed to it until I happened to look up and see her. I'm happy to say that Avalon has "the eye"!

Applause again rippled through the group, this time stronger and with some conviction. Avalon hunched her shoulders and drew her elbows tight to her sides, but smiled shyly and widely. In the moment after the applause she said, "I only saw a bit of bone. Julie found out what it was."

Michelle hugged her. Sometimes her daughter did understand other human beings. Sometimes she had had to learn the hard way. When in first grade a language teacher came in twice a week to teach Spanish words, Avalon asked Michelle to bring some Spanish books for kids from the library, but then she made many other students angry by asking why they were learning so slowly or by speaking to them in Spanish they could not understand. By comparing herself to other children and most adults, Avalon came to realize that not only did she learn languages unusually fast, but she recognized large and complicated patterns, almost like pictures coming into focus that were just a blur to her classmates. The teacher had called Michelle to come in for a parent-teacher conference and asked her to explain to Avalon that the other children were "learning as fast as they could."

At the after-supper briefing Dr. Art announced that as soon as he and the mapper had laid out a grid of squares in the morning, everyone should concentrate work in the area around the skull and arm bone. When each digger had completed work in his or her level of the sampling squares, they should report for assignment to one of the new squares around Cooler's discovery.

During the next and final week of the dig, Michelle and Dr. Art joined the diggers, kneeling side by side and taking turns at the sifters that screened the dirt. Dr. Art even enlisted Avalon for an assignment, which was to use her sharp eyes to watch for smaller fragments they might have missed. The girl had quickly impressed him with an unusual—no, a truly rare—ability to see small detail and distinctions between objects. The group's finds quickly grew into an odd collection of bones and fragments of both baboons and hominids, but nothing linked them together as clearly as Cooler's skull and its prey. The exception was the top of a small human skull, in which Avalon was first to spot parallel scratches of baboon fangs in the temple. "You're right!" Dr. Art said, then added his conditional, "At least if they're not rodent or stone scratches. I don't think they are, but. . ."

"But what?" Avalon asked. A couple of diggers laughed and rolled their eyes.
"Just but. Life is full of buts and ifs and maybes," Adamski said.

"Well, maybe," Julie replied. Even Adamski laughed.

All their finds came from the same stratum, both species mixed together and no other species except the occasional small rodent bone. Burrowing animals, mostly rodents, plagued most archeological sites. Adamski sometimes referred to the dubious location of an artifact or marks on it as "the work of time travelers, maybe." A few days ago he had said this while Avalon was listening and she declared confidently, "I want to be a time traveler. I think it's possible." When Avalon had gone to bed, one of the two married women asked Michelle, "What are you feeding that girl?"

The simplest explanation for the mix of human and baboon bones that issued from every day's random chatter was that a group of hominids trying to ford the river had been set upon by baboons and took a few of the attackers down into the mud during the struggle before they perished. Almost as if they were playing video games without a computer, the young diggers engaged in re-creations of *Homo sapiens* vs *Papio*. No one was sure which species of *Papio* baboons lived in the area, but like modern baboons, this *Papio* tribe had been strong, with teeth much sharper than most primates, and it had traveled in large packs. If they had been like modern baboons, the packs could have numbered over 100, led by aggressive males much larger than the females.

Would a band of *Homo sapiens* even have dared cross a river near a pack of baboons? One digger said, "It would have made the Normandy invasion seem simple."

"But our guys were much smarter," another digger added.

"Not smart enough apparently," Cooler chimed in. "Not then anyway."

Adamski decided to register a small box of bone fragments and teeth with local authorities, receiving permission to take them out of the country. Michelle suggested that Julie and Art could talk to her colleague in Oregon, Donald Koskin, about the find, asking his advice and perhaps even entering a collaboration. Koskin was a primate specialist at the University of Oregon in Eugene, Michelle's home campus. He had the tools to do a DNA analysis similar to those that revealed the Neanderthal genome. Michelle suggested that Julie might even be able to use this for an honors paper, though it was really the subject of doctorate work. Julie said to herself, *Dr. Flanagan, I presume?*

At the closing party Dr. Art thanked everyone for contributing to a historic find, "assuming the lab work and our peers agree," he added. An accordion player named Dave from Duke University sang "Ballad of the Baboon Battle," accompanying himself on the small squeeze box he had managed to lug to the site. The ballad's heroes were the alpha baboon Papio (sung as pa PIE yo and originating with the baboon genus name *papio*) and the *Homo sapiens* matron Mama Mia. Everyone joined in the chorus:

Papio, Papio if only you had known
That tasty Homo sapiens
Had many more neurons.
You were just empiric,
Your victories were pyrrhic.
We'd soon be back to see ya,
Thanks to Mama Mia.

Chapter 3

January, 2020

The two strangers in the lecture room listened and watched carefully as Professor Murphy clicked the slide advance button, and her last image appeared on the screen, seeming to be little more than the arid soil and stony outcrops of any desert in the world. "So," she continued, "How can we understand what Avalon and Julie found?" She paused very briefly and glanced at the two strangers who looked back, revealing nothing but polite interest. Michelle continued, "This is a photo Dr. Adamski took at the dig. Can you see anything? No? Well, now you know why people say that someone has the eye."

Michelle used her laser pointer to outline what looked to be just another rounded, brown stone. "Let's look closer." She pressed the clicker and the rounded stone resolved into the top of a partially buried skull. The next slide showed the skull cleared of debris so that the upper jaw was visible. "Do you see those teeth? Julie brought one back and showed it to Professor Koskin here on campus, and that's how we discovered Ma."

Those two in the back, Michelle thought—evangelicals, or something like that. Creationists. So, what could she do? Nothing. Whatever they were up to, she would continue to describe the origin of humanity, and it wasn't Adam and Eve. It was Ma.

Chapter 4

160,000 B.P.

Her name was Ma, the sound that she made to call attention to herself. In Ma's genome was a mutation that had given her an intelligence no other member of the clan had ever experienced: in her mind she could sometimes see ahead in time. She also saw and felt the past not just when she reacted in fear or pleasure, but when she wanted to see it and feel it again. Millennia later, when words could be used to describe her abilities, the words would be imagination and memory.

One of Ma's clearest memories was the time she first imagined the future and knew what she could do. The clan had found new territory with plentiful seed grasses, and the males had killed a young zebra that they found with a broken leg. Everyone had feasted for 2 days. In the evening with the sun setting they were lazy in the last warmth of the day.

Ma, still a girl, but in puberty, had left the gathering of women and now sat alone on a small hillock at the edge of the group. She was watching the clan's leader, a large and powerful male who had both hands above his head gripping the horizontal branch of a tree, with half his weight supported there and half on his feet. Behind him a female carefully worked through the long hair on his head, shoulders and back, picking out ticks or other parasites, eating those that were choice. As Ma watched she noticed that his penis grew and rose in front of him like a thick stick. The female grooming him also noticed. She reached down and touched it. His body tensed. She groomed for a few more seconds and touched it again, then slipped in front of him on knees and hands and elbows presenting her bottom, ready to take his thrust. He leaned over her and she arched to take the weight of his much larger and heavier body. Holding her shoulders and making a deep noise that came from his chest, he began to thrust, made a loud long throaty noise and slowly rolled off to one side. The female looked back at him, stood up and joined several others who had been watching.

Ma had seen this often but it had never interested her as it did now. She felt a contraction of the skin around her nipples and the nipples themselves hardened and tingled. Between her thighs she felt very warm and wet. Her first menstrual blood had begun to flow. She grabbed a handful of dry grass and wiped herself as the other females did. She knew what she wanted and she knew who she wanted.

She also knew that other females desired the male she had just watched. She had seen them competing for his attention, blocking his path, standing up straight and stretching for his inspection, grooming when he permitted. Into her head came a dream, only she was not sleeping. She could not know it, but she was the only one of her tribe who could dream standing up with her eyes open.

For several days afterwards she imagined pictures of herself and the male she wanted. He began to come to her mind with a sound—Om, Om, Om—and that became her name for him, the sound she heard in her head whenever she saw him or thought him. She also had begun to think other things. She could think about the sun, the stars, certain brightly colored fruits and flowers, and when she thought these things, she felt at ease, content.

Often she thought of Om and herself, but always she knew that the larger, stronger females would crowd her away, even bite if she approached him. Om had to choose her. The answer came from a bird. The bird was gray with a brown back and white tail feathers. Several times it had flown past her, always the same direction. She followed one afternoon, and found it probing inside a hollow tree. When bees buzzed out Ma knew at once what was in the tree. She had tasted honey in its comb only a few times and only a few drops because the clan fought over it, and she was not strong enough to win a piece. Bees she could fight, and she would because she had imagined herself giving this honeycomb to Om.

The bird flew away at her approach. She knew what bee stings felt like, but her purpose was stronger than her fear. She walked to the tree. Bees buzzed angrily around her, began crawling in her hair, and she felt stings start on her shoulders and breasts. Her hand inside the hive felt on fire with their stings, but she touched the comb, ripped out a long wide heavy piece and ran from the tree brushing away bees until no more clung to her except those that were dead and dying. She stopped to pick off the dead bees, and finally licked the honey that was dripping from the comb.

She wrapped the comb in leaves to hide it. She would have only a few minutes among the others before they detected her prize by its rich, sweet odor, so before she was close enough for anyone to smell or to look closely, she stopped and located Om. As soon as they were in positions where she could reach him before any of the others might cross her path, she walked quickly to him, pulled the leaves off the strip of comb and held it out to him. No one would dare challenge his right to this delicacy. He cocked his head to one side and glanced at it for only a second before taking it and beginning to eat. As he ate he paused several times to stare at her. When he had all of it in his mouth and was chewing the last of the honey from the compressed wax, she held up her honey smeared hand before him. He reached out and took it with a firm but not an abusive grip, and he licked the honey from her palm and fingers. When he was done he stared once more at her for several seconds, not harsh, not dismissive. She looked back into his eyes, something few males or females dared to do. Then she smiled. He nodded, turned and walked away.

A few days later the big male hunkered in the clearing of the veldt and watched her with his head cocked to one side. He was puzzled, and studied her. She had a familiar irresistible smell. Several other males were watching her too, but they waited behind him just out of sight in the tall grass. He studied her because she was different. Differences unnerved him. Differences often signaled danger, but this female was different in a way that was interesting.

First, she had done what no female had done—brought him honeycomb. In fact, no one brought anything to him voluntarily. When food was enough for all, everyone could take something to eat, in turn of course, in proper deference to more dominant members. Only mates shared and mothers shared with children.

When food was scarce Om could always take what he wanted from anyone, and he often did. And so on down the hierarchy. The only resort the lower had against the higher was to hide what they had or to leave the clan. Few who left ever came back, because hominid clans survived by living in bands, hunting and gathering together. Even so none of them died of old age. Most were eaten or died of a wound or disease. A lone hominid was easy prey for leopards, baboons, chimps, and hyenas. They lived longer inside the clan, but only by each accepting his or her place, and only until injury or infection or age meant they could not participate in defense or keep up with the moving band. Mothers would leave sick children. Children would leave ailing parents. This was the way of their clan in which they were born, lived and died. It was the way of most of life in their world. It was not complicated. It worked to eliminate the weakest and perpetuate the strongest and most able.

But Ma was different. Her behavior disturbed Om. She had the same dark purple-brown color, but something about her movement was different. He looked at her several ways but he could not understand why she seemed different. She tended not to mix with the other females, but often played with their children, almost like a child herself, making sounds they listened to but could not imitate. Older females did not play with children, but they had learned to follow the young female when she made a buzzing sound because it meant bees and honey.

Most unnerving to the big male, when he looked at Ma she did not look away—she looked back and she showed no deference or fear. She was not looking toward him now so he stood up and barked, "Hrrgh!" She turned to face him straight on, throwing out her chest to display prominent breasts. This made him excited, angry. He barked again, but she was not frightened. In fact, her mouth widened, her eyes narrowed and she made a soft sound, "Om." She said it again as she leaned a little forward, and in a slow swinging stride that came from being the fastest of the female runners, she came to him.

She stood so close that her breasts touched his chest. She kept looking into his eyes, and he looked back, puzzled. His advantage in height and power always made others, both male and female, look away. She continued to look up at him but unlike the way any other looked at him. Her look was soft, her eyes wide, her mouth wide and turned up at the corners. Mothers looked at their children like this. He was at first angry but increasingly curious about this female who would not look away. He felt his arousal stiffening and it brushed her groin. He waited for her to turn and present herself but instead she put her hands on his shoulders and pulled gently downward. Slowly he followed her down until he was on top of her as he had never been with another. She lifted her legs around him and he entered. He finished quickly but when he tried to push himself up she held on to him and looked into his eyes again. "Om," she said. Still holding him she said it again, and again.

From that day on she could always get his attention by saying his name, Om. He continued to mate with other females, but only with Ma did he lie face to face. Only with her did he rest and stay inside her until she released him.

Over the next 4 years Ma bore four children by Om. She also achieved a status almost equal to his; in fact, in some ways her following was stronger because she won it by respect alone without threat or force. The others feared Om but admired Ma. She had begun using the sound Ma for herself because it had been the sound her first baby made when it wanted to nurse. She began to call attention to herself using that sound, and very quickly the others had learned to call her attention with the same sound. And so a name had been born. They also learned the name Om, but they did not dare to use it in his presence any more than they dared to stare him in the eye.

She had first become a leader in foraging. She was almost always the first to find a stand of grasses with edible seeds. She knew better than anyone where to dig for tubers. The sounds she made became the names of the plants and animals they gathered. They no longer had to see each other to know where food could be found. They didn't even need to see the food. They had language, but only Ma's different brain could hold a word and its image in her mind more than a second. Only Ma could think. Millenia would pass before any of her species would think about thinking, but it was thought and remembered thoughts that made Ma a leader, and it was her leadership that gave them the advantages needed to survive in a world where food was increasingly scarce and competition increasingly brutal.

By creating names for things, she could also draw the clan's attention to enemies who were not present or who had threatened before. More than once she had saved a clan member from a deadly snake or crocodile by making its sound into a word the others understood. She could do little to save them from the stealth of hunting lions, but the cats usually picked on the sick and the old, or children who wandered too far from the tribe. Such deaths came one at a time and were relatively few.

Much greater danger came from the chimpanzees and baboons. She called the baboons Ka after their coughing sound. The males were savage and strong and wildly aggressive. They moved faster than Ma's people, and they climbed better and they had powerful hands that could also grab and hold while their fangs and powerful jaws did the killing.

In the past Ma's people had survived mainly by scattering and by numbers. The baboons were more eager to eat than to fight, so as soon as they had subdued enough hominids, their hunger set them to eating while the rest of the prey escaped. Nevertheless, many hominid groups were wiped out if they were few in number or suffered several attacks in quick succession.

After Ma's clan had lost more than a dozen members to a baboon attack, Ma worked out a defense. Using her name for baboons—Ka—she held her clan's attention while she herded the females and children into a tight group. She gave several of them rocks she gathered and soon all of them were cradling rocks in their arms. They already knew the power of stones to break other stones, crush bones, cut through hides, and soften tubers and mash seeds. They had thrown stones during attacks, but always at random. Together they became a battery of stone throwers with stored ammunition. Using Om who had come to appreciate her examples, Ma arranged the males around the inner circle. These males knew how to use heavy bones as hammers and clubs and how to use long sticks to knock down birds' nests and fruits. They quickly understood her instructions for defending the clan and themselves with these weapons. They learned that the baboons and chimps, not to mention hyenas and cats, saw such a tight, armed group as something like a single dangerously equipped beast.

Success bred repetition and repetition brought a dim understanding. The females seldom traveled without a stone, the males without a club or stick. Instead of scattering in fear when attacked, they now gathered together in defense and offense. Chimpanzees did the same, but Ma's clan could walk and run on 2 ft. The arms that had evolved when primates were tree dwellers were now free to swing clubs and throw stones. While other hominid clans disappeared, Ma and Om's clan survived and slowly grew. But as they grew famine became their greatest enemy.

Food was scarce. Neither Ma nor anyone in the clan knew the climate in their part of Africa was becoming steadily drier, but she had absorbed the fact that they had usually been most successful in finding new food sources by traveling in a certain direction. In fresh territory the males who hunted in packs were more likely to find prey animals. Each morning when she decided that the clan should move on, Ma would wait for sunrise, then begin to walk, the others following, keeping the sunlight always on the arm that she used most often to throw stones. The clan would find shade when the sun was highest, then forage for food and look for shelter in the cooler late afternoon. The result was that without knowing their location, without having a purpose, for several years Ma had been gradually leading them north.

With success came recognition, at least for Ma. Whenever food became scarce where they were, she would stand tall, sniff the air for long minutes, sometimes for hours, wait for the morning sun and begin to walk, making the same sound that she always made when she found food. The others had learned that sound and understood its importance.

Between places where food was enough to sustain them they wandered for many days across barren lands with dead and dying trees and brush where they continually lost weaker members to thirst and hunger and to the cats and hyenas always ready to pick off stragglers. Just as often they were the ones able to surround or chase down other animals, stoning and beating them to death. Sometimes they survived on bark, leaves and roots.

Like baboons and other primates, the clan also sometimes found water by digging in old river beds or near dry water holes. Ma, however, had a much sharper sense of where water lay. She could read the signs in certain combinations of land forms and vegetation. When they were particularly thirsty she led the others in turning over rocks and licking the condensation from the bottoms. Under the rocks they also found worms and grubs.

One day when her fourth baby was still clinging to her neck they had been hungry and thirsty for several days. They had survived by licking dew from rocks, sucking dry an occasional grub. For a month they had been losing numbers to predators, exhaustion, and dehydration.

They were moving very slowly now, resting often in whatever shade they could find. At night when the arid land was most alive with predators, they gathered in a tight group dozing, sleeping, waiting. They did not fear the familiar sounds no matter how dangerous the maker. No predator made noise when it hunted. They feared the silence, and the moonless silence most of all. When they sensed the approach of a predator or saw a shadow in the darkness, those who were awake set up the cry and woke the others and together they made a great noise and banged rocks and stones and bone and stick clubs. Their numbers and vigilance was their safety, but their numbers had been shrinking and fatigue dulled their senses and sleep reduced their vigilance.

Although this clan had always been wary of other hominid clans, even those most like their own, that was one worry they no longer had. For a long time they had been traveling across land where they saw none like themselves, only an occasional small group of short statured hominids with faces like monkeys, and these ran away.

One afternoon several members stopped to sniff the air and made a pleased rough "Hmmm, hmmm." Ma said, "Gul," and sniffed and nodded her head and again said, "Gul," her sound for water. They knew the sound and passed it among their scattered numbers. When they climbed a very low ridge of broken stone and low brush they saw the green rushes and low trees along the remains of a river that ran broad and very shallow across the land. And before they saw the water they heard the baboons.

"Ka, ka, ka," Ma said quietly. The word rippled through the clan.

The clan's adults looked down to the river uneasily. The baboons were relaxed, many of them feeding. The light hot wind was blowing up the slope away from the baboons. They did not smell the watchers. Om paced back and forth between his clan and the ridge, keeping them out of sight and dampening the more aggressive males whose caution was all but burned away by the food and water so close at hand. Ma saw the way. To Om and two or three other males she pointed out a couple of baboons sitting a short distance downstream from the others with their backs to the ridge. With gestures she pantomimed her plan. Om and several others would take them by surprise with spears and stones. The others would panic, and at that moment the rest of the clan's adults would come pouring down the hill screaming loudly, throwing rocks, waving clubs and attacking any baboons who did not flee. The plan relied on surprise and fear that would not allow the baboons time to understand the odds.

Om gathered a band of males and made sure all had strong sharp stick spears. Several also had clubs. Ma watched as they made their way toward the outliers hiding behind rocks, then launching themselves onto the pair with loud shouts and whoops, felling the two baboons and raining blows on them and piercing them with the spears. At the first shouts the rest of the clan set up their own battle cry and ran down the hill toward the rest of the baboon pack, chasing them into the river and across it and into the hills beyond. Two or three of the baboons that had been too sated or too old or too panicked to take flight in time lay in the water which flowed red with their blood. One of them was still clutching what looked like an arm in her jaws when her heart was pierced by Om's sharp spear.

The clan spent little time over the dead baboons but fell down to quench their thirst, and then to look at the flesh that had been left behind. The first to look at this flesh began to make the grunting of confusion and fear. Ma too looked at the remains of a body, an arm and a hand just like hers.

Whatever hominids had been killed by the baboons, they had not been from Ma's clan. They too had been moving north, but they had been too few. Although night was almost upon them, Om and Ma had no trouble in gathering the clan and moving off in a direction that would take them away from the baboons, at least from this group.

Over the next months and years Ma learned to lead and became at ease leading. In the mountains finding water had become easier and often they found caves with safe shelter. Most important they came to depend on their weapons, and were never far from them. For Ma, however, leadership was an uncomfortable status. She was very much one of the clan, but also keenly aware that she was different. So were her children. With them her language and theirs rapidly grew. By the time she felt herself becoming tired and noticeably slower than the rest of the clan, she and Om had had many children, a family that dominated all others in both size and understanding. All the children thought.

They had words for everything important in their world and a few words for feeling and judgment. AaaGA meant yes. Nye meant no or bad. At times as they rested and groomed each other the sounds and words they used became chants.

Their slow journey led them across desert lands to the shores of a large lake where they caught fish in the shallows and found birds nesting in the reeds. As usual where food was abundant, they had many competitors, and some of their own number became food for others. The abundant crocodiles that hid in the mud and shallow waters sometimes ambushed them by propelling themselves out of the water with a great roaring hiss. The larger crocodiles could also outrun a hominid in the marshy ground and thick reeds. They took several clan members. Om was the last of these to be taken as he stood mired in knee deep mud feeling for fish or turtles.

The lake waters tasted of salt and the clan moved steadily north along the west side of the long water toward distant hills. One of the first sons that had been born to Om and Ma, now in the prime of young strength became the clan's leader. He was a good word maker like his mother. He was impatient with those who could not make words or understand them.

At the north end of the lake they followed a river upstream into higher and cooler country where rains fell more often. Green hills covered with trees descended to the river valley. With very little travel they could gather plant food along the river, chase fish in the shallows, gather fruits and nuts from the trees and berries from bushes.

The grandchildren of Om and Ma multiplied here, and the easier life favored longer life. Ma lived to see the sons and daughters of her children and even a generation after these. One morning she did not wake up. A quiet death like hers very rarely occurred, and for a while her children and grandchildren tried to wake her and hunkered nearby waiting for her to stir. As soon as one of them began to repeat her name, the others joined. They wove her name into other words they knew and sounds they made and for several hours they sat around making that mournful chant. Then one of her older daughters began to dig with a sharpened stone they used for grubbing tubers. Others seized stones and dug and they continued their chant until the body was deep in the hole, then covered with dirt and stones.

Michelle Murphy came to the end of her lecture. She fixed her eyes on the young man and the solid woman in the back of the room. Their faces showed what appeared to be genuine interest, even admiration. She looked down at Julie and

Avalon in the front row. Julie gave her a reassuring nod. Avalon's eyes were closed, but Michelle recognized her thinking posture. She didn't have time to wonder what her daughter was thinking.

"Ma's clan grew and divided. Some of her children continued moving north through the mountains and down into the valley of a new river and along that ever widening river to a great delta and desert that rimmed a vast sea across whose waters they saw only sky. And their descendants with no idea where and from whom they had come, thought and spoke and chanted among themselves as if their kind had always done so. 100,000 years after Ma's death the first small groups of humans began to leave Africa, crossing into a fertile region that would one day be called the cradle of civilization. The pioneering humans were followed by many others, some venturing toward the rising sun, others toward the setting sun. Several species of hominids were already roaming in small tribes in the lands we now call Europe and Asia. They included Neanderthals and the three foot high Homo floresiensis of Indonesia and the widespread Homo erectus, a small but well traveled hominid with a skull much smaller than Homo sapiens. These species might one day have developed the intelligence and technology to populate the world. but instead they either melted into the clans of advancing Homo sapiens or were destroyed by them. For the next 50,000 years all the billions of humans from Iceland to Hawaii, from Siberia to Tierra del Fuego would be the descendants of Homo sapiens and would carry Ma's mitochondrial genome."

"No one knew this or believed it possible until 1987, when Dr. Rebecca Cann and her colleagues at the University of California published a paper explaining how analyses of human genes led inevitably to this conclusion. Using new advances in laboratory analysis, they examined the DNA present in mitochondria and passed from one generation to the next only in the ova of females. Cann and her colleagues studied mitochondrial DNA from people of many ethnic backgrounds around the world. From the differences and similarities in the DNA, the team could estimate how long it took for the variations to evolve, and could trace the route backward in time. The trail led them to a conclusion that others would confirm. Within the mitochondria of all humans were base sequences in the mitochondrial DNA of an African woman who lived some 200,000 years ago. The long forgotten woman buried by a river in southern Ethiopia was once more given a name—Mitochondrial Eve."

The two strangers in the back of the room joined the applause as Michelle waited for the overhead lights to come on. She took a long drink of water and thanked the audience for their applause. She had not expected the strangers to leave, and when she looked at them, the woman had nodded slightly in acknowledgement and with the clear message, *We'll wait.* Michelle shrugged inwardly. Since they made no move to go, she assumed they had something to say to her, and she'd find out who they were and what they wanted.

Meanwhile, she took pleasure in knowing that her talk had been well received by an audience of knowledgeable students and professors who were keenly interested in the science. A few faculty members might know that this was her first public lecture since the death of her husband Hank 6 years before in Indiana. He had been a rising star in archeology and ancient art. Even fewer knew that Michelle had once been an astrobiologist dedicated to tracing the elements of the first living cells to their origins in other parts of the universe. Only Michelle and one colleague in the audience knew why she had left astrobiology for genetics, and only he knew that she had chosen the subject and the timing of this lecture to honor Hank's memory and to resolve a disagreement that had stood between them for months before his death.

That disagreement had begun a year before Hank's death and it was about Avalon, who now stood in the front row looking up at her intently, on the verge of asking a question. Michelle thought she knew what it would be and was not ready for it. She called down to her daughter and Julie, "See you two later."

Avalon paused for a second, then nodded acceptance but a nod angled to one shoulder in skepticism. She followed Julie up the aisle to the exit.

Chapter 5

Michelle unplugged and stowed her laptop while the lecture hall was emptying, then looked up to see the two visitors wending their way through the chattering students to the front of the room. The woman smiled pleasantly and said, "I apologize if we were a distraction." Michelle straightened up and was about to say no problem, but before she could speak the woman introduced herself and her companion. "I'm Rhonda Grable from the U.S. Department of State. I work at the Central Asia desk." She looked up at her tall colleague. "This is Evan Lipkovich, from the Central Intelligence Agency." They both held out thin wallets with photo identification accompanied by the offering of a formal business card with department and agency seals. Michelle could hardly believe they were pretending, so she put the business cards in her briefcase, then looked at each wallet and handed them back. She wanted to say, "How would I know if they are real?" but she didn't. She said, "Obviously I'm curious. I doubt that you're here to learn about the *Homo sapiens* and baboons. What brings you to Eugene?"

"Can we sit down somewhere and talk, maybe your office?" Grable asked.

Michelle thought for a moment, checking her "to do" list for the rest of the day. "Sure, I have a few minutes. This way." As they followed her through the side door exit, Michelle was slightly amused to be leading the State

Department and the CIA to her humble office. She stopped just outside the auditorium where a small stand sold breakfast and lunch items. "I always get coffee here after lecturing. Would you like something?"

"Oh yes," Grable said. "Just what the doctor ordered. Or maybe the professor. My treat."

A few minutes later, they carried a double latte, a cappuccino and a black coffee into Michelle's small, austere office. She invited them to sit and pulled her own chair from behind her desk as she did in student conferences. When they were comfortable, Grable said, "Evan, it's your turn."

Lipkovich nodded and looked directly at Michelle. "Professor Murphy, you will receive—maybe next week—an invitation to be a keynote speaker at a conference on genetics and intelligence."

Michelle rapidly processed the possibilities, but none of them made sense. "A conference organized by the State Department and the CIA?"

"No, no" Lipkovich said almost apologetically. "The invitation will come from The Kazakh Institute for Biology of the Future, or for short the 'Asfendiayarov' Institute. The title of your talk will be Transcription Factors in Neural Development and Primate Evolution."

Michelle raised her eyebrows. This was getting weird. In her mind Kazakhstan was just a shapeless blob on the map of central Asia. "*If* I get the invitation. *If* I accept." She stressed the 'ifs'.

Grable said, "We hope you will accept. That's why we're here. We think you're the right person."

"Well, I can't be humble and pretend I don't know the subject pretty well. But I suppose you know that already."

"We do," said Lipkovich.

"I have to admit," Michelle said, "I'm not wildly enthusiastic about all the personal information the government seems to collect these days."

"We understand your concern," Grable said, "But all we know about you is on your faculty web site." Quite clearly the way she spoke said, *let's not be distracted by that debate.*

Lipkovich added, "To be clear, we know nothing about you that is not public information. We would like the benefit of your expertise as a scientist. We're not asking you to be a spy or recruit anyone or to do anything you wouldn't normally do at this kind of conference."

"But you didn't come all this way just to tell me about the invitation and encourage me to attend."

"Actually, we did," Lipkovich said, "Because what you might learn could be important."

"I think it's time to say more precisely what I can do for you."

"Yes, but before we go further, everything I tell you must be kept confidential. If you agree, I need you to sign a promise of confidentiality."

Michelle said nothing as she thought about Lipkovich's request. She was aware that Grable was looking at her.

"So if I want to satisfy my curiosity, I have to sign," Michelle said.

"You can put it that way," Lipkovich agreed. He had the slight grin of a boy who had gone fishing and saw his bobber go under. "You don't have to sign. You'll get the invitation anyway, and whether you go or not is up to you, of course."

Lipkovich drew a sheet of paper from his briefcase. "I hope you will sign. Then I can explain why we're here." The declaration was simple—as an American citizen, she would be required to keep everything she would say and hear to herself, including the identity or affiliation of Evan Lipkovich.

"I'm a scientist," Michelle said. "I try to discover and publish new knowledge, not keep it secret."

Grable smiled slightly, sympathetically, or at least Michelle thought so. Lipkovich waited. She read the paper once more quickly before signing and handing it back. Lipkovich thanked her and put the paper back in a folder in his briefcase. Michelle realized she had just agreed to keep secret something that she knew nothing about.

Grable nodded at Evan and said, "Go ahead."

The agent looked at Michelle for a brief moment of assessment, then continued when he had her full attention. "Let's start with what must be on your mind, why the CIA is involved. In Central Asia my official role is with the State Department as a roving scientific and cultural attaché for all our embassies in the stans. The stans . . ."

"Six stans," Michelle said to save him the trouble and time of explaining.

"Yes, but Kazakhstan, not Afghanistan or Turkmenistan. The next thing I'm going to tell you will seem a bit silly, but bear with me." He looked at Michelle, making sure she was listening. "Your invitation will come from Arman Akenov, the director of the Asfendiayarov Institute."

Michelle searched her memory but nothing came up. "Never heard of him."

"If you followed international horse racing, you would have. Akenov doesn't just race horses, he is regarded as a genius in the arena of horse breeding. His horses are winning races in Europe with times that set world records, not just by a nose but by several lengths. He is a national hero in Kazakhstan, and has a position in government as science advisor to the president."

Michelle shrugged. "I know nothing about racing or breeding horses. I don't see what this has to do with me, or why he would invite me to give a talk." Her mind, however, was already making associations—breeding, genetics—and she did know something about genetics.

"That's an interesting question," Lipkovich said. "Let me tell you why."

Michelle sat up straighter in her chair, indicating that she was ready to listen.

Lipkovich leaned forward and lowered his voice, almost as though he assumed someone outside the room was listening. "Homeland Security means we not only need to know about existing threats and weapons, but to anticipate new ones."

Michelle couldn't help smiling. "A Kazakh horse breeder is a threat? I assume Akenov is not a mad scientist."

Grable said, "Angry scientist, maybe, and they are more dangerous, particularly in the former Soviet Union. They will work for the wrong people."

Lipkovich continued. "Mad scientists are few and far between. They never have an organization or the kind of funding that gives them power. But angry scientists who attach themselves to a cause, a government, or a religion have both power and organization. Akenov has strong funding from his government."

"So do all the best scientists in America—well, directly or indirectly. Why is Akenov different?"

"We would be glad if he were not different," Lipkovich said, "But we suspect he is different, and if we're right, the work he does could be very dangerous in the wrong hands."

Michelle took a sip from her coffee, put it down on the desk and waited.

"First, a little history lesson," Lipkovich said. "Have you heard of Vozrozhdeniya Island in the Aral Sea?" He pronounced the Russian name with what Michelle assumed was a native Russian familiarity. Then he translated. "Some people render the name Resurrection or Rebirth Island."

Michelle's cell phone sounded a few sprightly bars of music she knew as DNA Suite, a musical translation taken from the base sequence of the human insulin gene. The number on the screen was Avalon's. "My daughter. I should take the call." Before she clicked the talk button on the phone she quickly answered Lipkovich's question, "The Aral Sea, yes," Michelle said. "The island no." Into the phone she said, "Hi, Stargazer, what's happening?"

"Nothing much," Avalon replied. "I got home from school and I was lonely, so I thought I'd say hi. Who were those two people at your talk?"

Michelle knew loneliness well, and even before Avalon had started school she had seen how much trouble she had talking to other kids.

"Sweetie, I suppose you want to say more than hi, but I'm in a meeting with them just now. Let's talk when I get home."

Avalon answered Michelle with a rhyme, as she liked to do in a game she and Michelle often played. "We dance around in a ring and suppose, but the secret sits in the middle and knows."

Michelle laughed. "Robert Frost. I'll be home soon and we can have a mother-daughter gabfest."

Michelle worried that Avalon had so few friends. A 12 year old should not be lonely. Maybe she shouldn't be home by herself either, but Avalon was no ordinary 12 year old, and they did have good friends and neighbors next door. Then there was Julie, who had become very fond of Avalon, often dropping by to chat and play chess, a game Avalon always won, even when she coached her opponent or gave them back multiple moves. Julie had once told Michelle that the only thing she could do better than Avalon was to bake the oatmeal chocolate chip cookies they both loved.

Michelle turned back to her visitors and explained the phone call. "My daughter."

Grable asked, "Is her name really Stargazer?"

"She has a lot of names, depending on what interests her. Lately she's memorized all the constellations and has been reading the stories behind the names."

Both Grable and Lipkovich nodded appreciatively. Grable asked, "What was that about Robert Frost?"

"Avalon loves poetry. She knows poems by heart, and when she quotes one I'm supposed to guess who the author is." Michelle paused, considering whether to share something personal, then decided that she could trust Rhonda and Evan. "My husband and I adopted Avalon from an orphanage in Almaty, so I visited Kazakhstan at that time and know a little about the country. In her first few years with us, we began to notice that Avalon was remarkably intelligent, far beyond other children her age. For instance, as a two year old she was able to repeat a nursery rhyme after hearing it just once. Avalon once told me that the reason she loves poems is that she is unable to create them herself. I think her mind is so precise that she can't come up with the nuances and metaphors of poetry, even though she enjoys them immensely."

"Was Avalon the young girl sitting in the front row?" Lipkovich asked.

Michelle was surprised that Lipkovich had observed Avalon and guessed correctly about their relationship. Her estimation of the CIA agent went up a notch. "That's right," Michelle replied. "But let's get back to Resurrection Island."

Lipkovich nodded and continued his story. "During the Soviet era, the island was a major center for secretly developing and testing biological weapons. It was guarded by the military but run by an organization called

Biopreparat. Like most Soviet research centers, it also had a residential colony of scientists and their families. When the Soviet Union began to collapse, Moscow quickly dismantled biological weapons facilities like Vozhrozhdeniye because they had signed a treaty prohibiting such research but had never bothered to followed its regulations. They hid the evidence on the island, dousing thousands of pounds of anthrax and other weapons with bleach and burying the stuff in pits. They also bulldozed the labs on the island and abandoned the place very quickly."

A thought crossed Michelle's mind—that she was listening to a description of Avalon's homeland.

Lipkovich continued. "There are two consequences that worry us. First, they could not have killed all the organisms. Second, after the hasty abandonment they exercised very little control, so no one policed the island and it was easy to get to. For looters the island was a gold mine of truck parts, furniture, building materials, scrap metal of all sorts, even clothing. Lots of valuable stainless steel in the labs. Our ambassador got permission to inspect the site in 1995 and the team saw pillaging everywhere. By the time western countries negotiated an agreement to help clean up the site everything that had any use or resale value had been taken. That agreement was in 2002, ten years after the Russians abandoned the place."

"And we helped with the clean up?" Michelle asked.

Evan turned to Rhonda, who knew this part of the story, and she filled in the details. "The State Department had two de-militarization initiatives in Kazakhstan. Project Sapphire successfully negotiated the disposal of all of their enriched uranium but we found their inventory list showed 4 more kilograms than the amount we recovered. Bookkeeping error or theft—we don't know. The other mission was the cleanup on Resurrection Island."

Lipkovich picked up the thread. "We knew what was on the island because a Kazakh microbiologist and biological warfare expert defected. That was Colonel Kanat Alibekov. He calls himself Ken Alibek in the US—taught at George Mason University, then went on to push some dubious immune system supplements and operate a consulting firm. Alibekov told us they were working with bubonic plague, developed new and more virulent strains of smallpox and had possibly put the ebola genes in the smallpox virus. They did a lot of testing—open air testing on animals. To make matters worse, the Soviet irrigation system for its New Lands cotton program was drying up the Aral Sea. In the late 1990s people could wade to the island. By 2001 you could walk there on dry land."

Michelle shivered. She knew very well how pathogenic bacteria and viruses caused disease, and also how their genes could be manipulated to make them even more virulent. "And the connection to Akenov?" she asked.

Lipkovich continued. "We know that Dr. Akenov made several trips to the island when it was under Biopreparat. We don't know why. We know from various sources in Kazakhstan that he is not only working with gene splicing in horses and other animals, but that he has some kind of virus research program." Lipkovich paused to consider what he would say next. Michelle waited.

"Professor Murphy, let me ask for your professional opinion. Do you think it would be possible to develop a virus that could produce cancer?"

Michelle was astonished. "You think Akenov is developing a viral vector for cancer? Why would he do that?"

Lipkovich leaned forward a little. Michelle couldn't shake her sense of him as a football player, a very intelligent, sly quarterback. Even if football were her game, she knew she didn't want to be playing against this man, maybe not even on his team. But so far she had not agreed to join the team, only to listen to the recruitment pitch.

"Well, anthrax is old fashioned," Lipkovich said. "It works, but it's yesterday's terrorism. We know that Akenov worked very closely with certain Russians and that their specialty on the island was viruses. We know a lot more, but . . ."

Michelle decided that his impeccably neutral voice and face had to have been trained. "If you tell me, then you have to kill me."

A small smile drew back the corners of Lipkovich's mouth and eyes. "Something like that," he said.

Michelle noticed Grable looking up at the ceiling. If this were a play, she would be off stage. Michelle wondered if she should offer another cup of coffee, but decided not to. No interruptions. Nothing to delay the enlightenment she was waiting for.

Lipkovich leaned back in his chair, folded his arms and looked directly at Michelle. "We are concerned about biological weapons in the hands of terrorists, of course. From the 9–11 attack, we know that a few of their leaders are total nut cases who would not hesitate to plant an atomic bomb in Tel Aviv if they had one. Or Washington, for the matter. Our government accepts that as a possibility, but when Netanyahu gave his speech to Congress in 2014 he made it clear that it's a certainty in his mind."

Lipkovich paused for a moment, catching Grable's eye, then continued. "Knowing what I know, I'm afraid that I agree with Bibi's judgment, and that's why we're so interested in the uranium and the biological agents on the island. But now we have another concern. Terrorists use weapons of mass destruction and want the credit, but governments, with few exceptions, use assassination. Usually they seek anonymity, but sometimes they might want to be suspected. For instance, the 2006 assassination of the Russian KGB agent who defected to the British was carried out with radioactive polonium. It's easily detected and traced. Easy enough to guess why he was killed, but we can only guess who. We assume the 'who' in this case wanted to be suspected, basically a message for others who might be considering defection."

Michelle suggested a label, "Semi-anonymous selective terrorism."

Lipkovich chuckled, glancing at Grable. "I like that. SAST, a new acronym. Yes, you could call it that. Now suppose you wanted to assassinate someone with complete anonymity and not even have the death be known as an assassination." He paused to let Michelle take that in.

She thought of the obvious—heart attack, stroke, accidents, then said, "It would have to be something that people, even apparently healthy people, frequently die from and a weapon that can't be seen."

Lipkovich raised his eyebrows. "Professor Murphy, if you ever get tired of university life come to Washington."

Michelle smiled, then replied, "Would that be a step up or down?"

Lipkovich smiled back. "Point taken. But now I can finish my story and satisfy your curiosity. If you know anything about the old Soviet Union and the KGB, you probably know that the death certificates for many of their victims said cause of death was heart attack. Even then few people believed it. Today, especially if the assassination crosses borders, it's much more difficult to hide the real cause of death. If someone in London or Washington dies from Ebola, we can immediately narrow the sources and go hunting. But suppose something else was the lethal agent?"

"Do you have something in mind?" Michelle asked.

"Only because it has recently been ordered by Akenov's institute, and at the same time we know the Institute received a large payment from a Swiss bank account controlled by a questionable Russian charity. You probably know about SV40."

Michelle said, "Of course. Just to make sure we're on the same page, you're talking about the virus discovered in the monkey kidney cells used to make the first polio vaccines, Simian Virus 40."

"That's right," Lipkovich nodded. "There's a lot of debate about whether vaccinations contaminated with it ever caused cancer in humans, but apparently there is no controversy that it can cause cancer. Would you agree with that?"

"How do you know Akenov has this virus?"

Lipkovich drew a long breath, closed his eyes a moment to ready his next play. "Can I just leave it that we know?"

"I'm sure you can," Michelle said. She leaned back in her chair and waited.

"I'm sorry," Lipkovich said. "Unless there's a reason you need to know, we prefer the rule that the fewer people who know something, the safer the source or the method." What Lipkovich could not say was that they had detected a suspicious parcel shipment via the red tape required for international delivery of biological hazards. A package with a large Styrofoam container and dry ice needs special handling and quick pickup. Their asset in Almaty watched from across the street when someone from Akenov's institute signed for the parcel at the FedEx office, then drove away in a black Mercedes SUV.

Michelle couldn't find any reason why she needed to know how the CIA traced the virus to Akenov. "I assume you know that my research focuses on the interaction between genes and viruses." She had not asked a question and neither Lipkovich nor Grable answered. "SV40 got in the news because it was in the polio vaccine, and of course the issue got oversimplified in the media with claims that it caused cancer. Let's just say it can do a lot of things in cells that may result in cancer."

"Thank, you," Lipkovich said. "I'm here talking to you because we think Dr. Akenov is developing this virus as a mutagen. Given the people he's worked with and other information we have, we think he's developing agents for biological assassination that can't be tracked. Indirect assassination is not new, but SAST with viruses is."

Michelle said, "So you want to know if Akenov is doing it or wants to do it?"

"We want an expert opinion, and that's your opinion. And we'd like to know a lot more about Akenov's work and connections."

"Why me? You could get a dozen people better than me."

"Maybe yes, maybe you're being modest. But the other people are not being invited to Dr. Akenov's conference this summer."

"Why do you think he will talk to me?" Michelle asked.

"Because you probably know more than he does about how viruses can alter genes. That would be a good reason to invite you to the conference."

"And you think he can't ask what I know without revealing what he knows and maybe why he wants to know more? This is starting to sound like one of those convoluted spy stories."

Lipkovich seemed to relax now and sat back with his hands clasped behind his head. "We think it will work that way, but we're not talking about what the public calls spying. All we want you to do is to engage in professional conversation with him, exchange ideas as scientists do. If you have to steer the conversation a little bit, do it carefully, make it natural, professional."

"You said there was a second reason for your interest in Akenov."

"Yes, there is. Akenov is a passionate Kazakh nationalist. Besides his work with horses, he is the director of an institute called The Good Shepherd Foundation. We have evidence that it's a front for a growing group of extreme nationalists outside of government."

"What does extreme nationalists mean?" Michelle asked. "Religious fanatics?"

"If you mean Islamic, no. No religion that's apparent in any case. It's not a group we know much about. For Akenov nationalism is an ethnic thing. He believes that certain groups of humans can become a superior race."

"Let me get this straight. You think Akenov is a Kazakh supremacist? This is getting way beyond anything I'm familiar with."

Grable answered, "All we're asking you to do is to use your powers of observation, look around, pay attention to who attends the conference, who's a scientist and who's not, that sort of thing."

"Suppose they arrest me as a CIA agent?" Michelle asked, only partly serious.

"First, you're not an agent," said Lipkovich. "There's nothing that would indicate you were an agent or anything like it. We're just asking you to tell us what you see and hear. But let's go back to this SV40 virus. I've been briefed on it, of course, but I'd like to hear what you think he could be doing with it."

Michelle was an experienced lecturer, so it just took a few minutes to describe how viruses can activate oncogenes and turn ordinary cells into malignant cancers. She also pointed out that in her research sponsored by TransTek, a local biotechnology company, she was developing state of the art procedures in which viral vectors would be used to repair missing genes in certain genetic diseases like hemophilia and cystic fibrosis. She was surprised by some the questions Lipkovich asked during her mini-lecture, as though he knew what she was doing at TransTek. Surprised, because she had a non-disclosure agreement with the company and everything she did was supposed to be secret and proprietary.

As Michelle finished up, Grable was shaking her head and had a wry smile on her mouth and a worry crease in her forehead. "And soon people like Akenov will be able to make their own highly targeted viruses?"

Michelle fixed her eyes on Grable and nodded. "People like me can do this. I am doing it in my work at TransTek, but my aim is to design viruses to repair cells, not make them malignant."

"But you're developing the same technology," Grable said.

"Yes, but I'm with the good guys," Michelle said firmly. She looked at Evan Lipkovich. The big man was rubbing his jaw, thinking hard. "Well," he said, "Like everything else, technology is the old two-edged sword with both good and bad applications. If you can do it, I'm sure Akenov will be able to soon."

Grable and Lipkovich exchanged looks. Grable tapped her watch, then turned to Michelle. "It's been very good meeting you, Professor Murphy. You know as much as we do now, and we have a plane to catch." Her visitors stood up, and Michelle did as well.

"I hope you will think about what we told you," Lipkovich said. "When the invitation arrives, let us know what you decide to do. You have my card, but don't use my email address. I'll call you on a secure line."

Almost as an aside as the two visitors were putting on their coats to leave, Lipkovich asked, "How good is the computer security at TransTek, would you say?"

Michelle was surprised. "What do you mean?"

"Given what you told us about the nature of your work and what TransTek does, there could be attempts to hack into the system."

Michelle thought about this for a moment as she watched them pick up their brief cases and pause by the door. "There are security guards at the gate 24/7 and all the critical areas in the buildings can only be entered with key codes. The computer system I don't know about except that I have codes and passwords to use."

Lipkovich said, "We'll have someone in NSA look into it."

Before she caught herself, Michelle blurted, "Look into it or hack into it?"

Lipkovich said a bit brusquely and with emphasis, "*Look* into it."

"Sorry, I didn't mean to be offensive," Michelle said, then added, "Well, I guess I was."

"It's okay," Lipkovich said. "We're used to it."

Michelle was back home in time to prepare supper. Avalon was in the living room, lying on the rug reading. When Michelle had hung up her coat and kicked off her shoes, Avalon announced matter-of-factly, "I'm going to be a geneticist. A biomolecular engineer I think."

Michelle stood at the door to the kitchen and after a moment's thought she said, "Should I be flattered?"

Without looking up from her book, Avalon said, "Maybe. I decided after your lecture today. So I've been reading."

"What's the book?" Michelle asked.

"Odd John," Avalon said. "By Olaf Stapleton. It's old, but really interesting."

Michelle recalled that she had read it too, years ago as a teenager. "Is there a particular reason you chose that book?"

"Yes, of course," Avalon said. "I think I'm a mutant. Like John in the book."

Michelle was stunned, and took a moment to gather her thoughts. "Okay. You are very smart. Much smarter than anyone else I know. But being smart, even being a genius—which maybe you are—doesn't make you a mutant."

Avalon rolled over onto her back and stared up at her mother. She liked most of her classmates but she had never felt close to anyone except her mother—this mother, and what seemed long ago, she had felt close to her adoptive father. She remembered that he often sat on her bed reading to her, sometimes looking into her eyes. "I think Dad knew I was a mutant," she said.

Michelle still stood at the kitchen door. She was not ready for this conversation. "Why do you think that?" she asked.

"Because he used to look at me for a long time without saying anything. He really wanted to know who I was. What I was."

Michelle had also pondered these questions, but she had hidden her thoughts successfully.

"Mom, he wanted to know what I want to know. Don't you want to know?"

Michelle walked over to Avalon, got down on the carpet and stretched out at her side. "Your father, Hank, he did want to know. I didn't want to know back then. It was too strange, and all I wanted was a daughter. But I think I'm ready now."

Avalon turned and put an arm around Michelle, hiding her face in the corner of Michelle's neck and shoulder, her black hair falling like a curtain across Michelle's face.

"Mom, don't be afraid. I won't leave you."

A few days later Michelle's cell phone rang. She pushed the talk button, but before she could say hello, she heard a voice she immediately recognized as Lipkovich, and then his face appeared on the screen.

"TransTek's system has been hacked."

"What? By whom?" Michelle asked.

"We don't know for sure, but our people say the hack didn't come from America or Europe."

"Kazakhstan?"

"Maybe, but we don't know for sure. You should inform the TransTek folks and have them beef up their internet security procedures. Meanwhile, don't take this as criticism, but if you do happen to keep any TransTek data on your laptop, you might want to remove it. Home computers, laptops, and even university computer systems are usually easy to hack into."

Michelle sighed. "It's a wonderful world, isn't it?"

Lipkovich answered with a few seconds of silence, then said, "I guess I should tell you that we also asked to have your personal computer checked to see if it was hacked. We found Powerpoint talks, various manuscripts, lots of Avalon images in iPhoto, thousands of personal emails, the usual things. But there was also a fairly sophisticated malware program that would let someone watch whenever you were using your office computer. It's best to leave the malware there, because removing it would be suspicious. Also, we can use it to send misleading information if necessary. Whoever put it there probably grabbed your user codes and passwords. I recommend that you continue using them for ordinary university and personal stuff, keeping in mind that someone is watching. For sensitive information like your TransTek work, you should buy a new dedicated laptop, set up new user codes and passwords, and only send encrypted material."

Lipkovich's image locked eyes with Michelle over 3000 miles between CIA headquarters in Virginia and Eugene. "Like you said, it's a wonderful world."

When she had slipped her cell phone back into its holster, she thought to herself, *It certainly is a world full of wonder. I see wonders every day and I have a daughter who makes me wonder.* Michelle had no idea what wonders her daughter might see in her own lifetime. She wanted nothing more than for Avalon to enjoy a long and happy life.

The invitation to the conference arrived ten days later in a large, bright white envelope bearing the blue and gold seal of the Republic of Kazakhstan. Beneath the seal was a logo she did not recognize and the word "Asfendiayarov" followed by what she assumed was an English translation: The Kazakh Institute for Biology of the Future. She extracted a glazed blue and gold folder with information about the Institute situated in the new capital city of Astana, Kazakhstan. Inside the folder she found a single sheet of the Institute's stationery. Sitting boldly in the center at the top of the cover letter were the sender's return address:

Arman Akenov, Ph.D.
Director of the Kazakh Institute for the Biology of the Future
"Asfendiayarov"
President of Good Shepherd Foundation
Astana, Republic of Kazakhstan

Dr. Akenov announced that it was his pleasure to invite "you my very dear Professor Murphy" to deliver a keynote talk for a session on Transcription Factors in Neural Development and Primate Evolution. "Your credentials in both the academic and commercial worlds make you uniquely qualified for this honor." The letter went on to give the dates in July, promised that all expenses were covered, including first class airfare, and that she would receive a $10,000 honorarium.

The amount surprised Michelle, because in the academic world it was reserved for Nobel Prize winners. The final paragraph said, "In addition to your unique scientific qualifications, we know from your biography that your daughter is Kazakh. If you are willing to accept our invitation, it is our hope that your daughter will wish to see the land where she was born and participate in conference related activities for youth sponsored and conducted by The Good Shepherd Foundation."

Chapter 6

2009 Bloomington, Indiana

Hank had earned his Ph.D. in archaeology and sociology by using behavioral economics to demonstrate why dealers, collectors, museums, and scholars all too willingly take part in the commerce of plundered, undocumented, and often fake antiquities that find places not only in museums and private collections but in scholarly dissertations that become the foundation stones of degrees and careers. He had been inspired by the work of his fraud-hunting hero at the Metropolitan Museum of Art. Oscar Muscarella had been the Met's Senior Curator of the Ancient Near East, the excavator of the Midas tomb and important sites in Turkey and Iran. Muscarella's books and lectures and unremitting criticism of plunderers and the dealers and collectors who finance them had attracted media attention that sent periodic tremors through the staid and often smug world of antiquities. Each book and lecture sliced through a new cross section of deception and corruption and fraud. Reputations withered in the light of disclosure, and occasionally a substantial tax deduction taken for donating a fake became a large fine. Muscarella's unrelenting and often sharply worded and witty crusade brought down upon himself the wrath of the Met's famous socialite director Thomas Hoving who tried to fire Muscarella. In an unprecedented legal move, the curator claimed tenure and, in a multi-year legal battle, he won. He stayed. Hoving left.

Like Muscarella, Hank's passionate honesty had attracted both strong enemies and devoted followers. One morning in June, he was up at 5 checking email. He had not been an early riser in the years before Avalon's adoption, but now his daughter took so much of the morning that he valued the quiet hours before Michelle rose and the two of them planned the day's agenda, mostly who would deliver Avalon to day care, who would pick her up in the afternoon, who would shop for groceries, all the duties life and love suddenly thrust upon parents when their first child arrives. The usual 50 or so emails were on his screen. He scanned through them deleting the spam and irrelevant campus announcements until just a few were left to read. One was from George Wilkins, his colleague at the University of Chicago who specialized in counterfeit art but kept an eye on antiquities sales. It read:

> Hi Hank—Just saw that Sotheby's is putting some gold and bronze sculptures into their auction next week. Check the online catalog—looks to be in your bailiwick.

Hank did check and saw that a collection of ornaments and jewelry, mostly bronze but one of gold, would be put up for bids. The provenance was unknown, most likely central Asia. From the photographs Hank was certain the collection would be Scythian, nearly 2500 years old. Similar objects had been found in the tomb of the Golden Man discovered near Almaty, Khazakhstan in 1969. Hank suspected that the artifacts had come from grave robbers during the Soviet era or even before, then sold from the plunderers to collectors who may have never known the origin or context of the finds. Throughout the Soviet period gold had always been an easy way to preserve and conceal wealth. Whatever the reason, these objects had now appeared at Sotheby's, and their history, if it had ever been known, would have been erased far back in the chain of owners.

When they had been in Kazakhstan for the adoption, they had visited a few museums. The lack of documentation had disappointed him. He had returned home to Indiana certain that the culture of corruption in the old Soviet Union together with the struggle to provide anything above life's basics had nourished a widespread black market in antiquities. Perhaps Kazakhstan deserved a little attention.

The gold ornament was the centerpiece of the collection, so Hank examined the photo carefully. The border was a beaten gold oval about 5 inches wide and over an inch high, and within the oval the artist had framed a horse galloping from left to right without a rider, forelegs extended with its tail streaming behind. Hank admired the craftsmanship. Something like this, if he could afford it, would be a wonderful present to Michelle on their anniversary, and eventually a keepsake for Avalon. However, a museum or a wealthy private collector would bid for it at several hundred thousand dollars. But he continued to study the piece. The horse's head with its wild eyes and streaming tail stirred something deep in his mind. He had seen this horse before. He walked to the bookshelves that filled one wall of the room and found a book he had bought in a shop in Almaty—Kimal Akishev's 1983 book *Ancient Gold of Kazakhstan*. With just a few minutes searching, he found what he was looking for.

"Ha!" he said aloud.

"Ha, what?"

Hank was startled and looked up to see Michelle standing by the door with a cup of coffee, still a little sleepy but smiling at him.

"Look at this," he said. "Look at the eyes." Hank took the book over to his desk and set it down in front of the computer screen.

Michelle came over and compared the photograph on the computer screen with the image in Akishev's book. "Same eyes, but the horse is running in the other direction."

Hank nodded. "The one in the book is the real thing, and if it were put up for sale it would be worth hundreds of thousands I think. The one on the screen will be auctioned at Sotheby's next week, and a collector or a museum will be bidding. But if it's a clever fake, which I think it might be, it would be worth little more than the weight of the gold."

"What are you going to do about it?"

"I'll warn Sotheby's of course, and have them send a note to whomever is putting it up for sale."

"Mr. Popularity," Michelle teased. "One day that kind of stuff will get you in trouble."

That afternoon Hank sent an email note from his work computer to one of the acquisitions specialists at Sotheby's whom he knew well from previous occasions when he had been brought in as a consultant. The expert, after some fast and intense research, sent his own note to the Swiss dealer who had acted as an intermediary for the owner of the items. That dealer in turn forwarded the note to Kazakhstan, where it landed on the desk of the director of an institute on the outskirts of Almaty.

Sotheby's web site later noted, "Item withdrawn from sale." Hank showed the notice to Michelle. She asked, "Don't you think somewhere a very rich person is pretty unhappy with you?"

"I think my friend at Sotheby's gets the blame, at least until I describe the fraud in an article, which I might just do since I'm getting interested in Central Asian material. After all, we already have one gem from Kazakhstan."

Hank never did publish a note on the ornament. A month later, with his article still little more than a few hasty notes, he received an email from an antiquities dealer in Switzerland requesting him to consult on yet another artifact. The dealer wrote, with apologies, that he had already taken the liberty of sending the item by FedEx, with return labels included. That was followed by another apology ending, "If I were not sure that this unique piece would interest you, I would not have sent it." The note concluded by saying that a FedEx shipment would arrive at his university address that day. As he drove to his office and parked in the faculty garage, Hank wondered what this was all about. Maybe another fake gold artifact? He found the FedEx package in his departmental mail box along with the usual assortment of campus mail. The large plastic envelope contained something that felt like a bulky box perhaps an inch thick. The department secretary had signed for it. He'd have to open the package to get the return label, so he decided to take it up to his office and open it carefully.

Sitting at his desk, Hank used the small blade on his pocket knife to slit the envelope and carefully remove a cardboard box. There was no note, and no return label unless it was inside. He used the knife to open one end of the box, which turned out to contain a sealed plastic bag with Styrofoam packing beads. The bag was heavy, definitely something metallic was concealed by the beads, and he could see what looked like a note and a shipping label. When the knife punctured the bag, he was startled by a hiss that released the beads along with what looked like powder puffing into the air and onto his hands. Had they gone so far as to pack it in some inert gas? Hank reached down into the beads, felt metal, and removed the artifact. When he saw what it was, he was surprised and disappointed. Surprised because it was a small, bronze swastika, perhaps 2 inches square, disappointed because it could not possibly be related to any known Scythian artifacts. Why would a knowledgeable Swiss dealer send him such a silly thing? He glanced at the note, which was just a single cryptic statement: "A token of appreciation for your services to the people of Kazakhstan."

Hank shrugged. Why indeed? Whoever sent this must be some sort of nut. He added the swastika to his growing collection of questionable antiquities and brushed the scattered beads into his waste basket, along with the box and envelope. For the rest of the morning he worked on a manuscript for a book chapter he had been invited to write, then put his computer into sleep mode and walked downstairs and toward the parking garage. He was looking forward to a lunch he and Michelle had planned to have at Samira, an elegant restaurant in downtown Bloomington that specialized in Afghan dishes.

As he pressed the unlock button on his car key, he heard the car beep, and simultaneously felt a deep throbbing ache in his thigh that brought him to his knees with a gasp. Then a pain shot down his left arm. He thought, *Oh God! Is this a heart attack?* Within seconds his vision faded to black as his mind shut down. There was no more pain, and he crumbled to the pavement next to his car. Another faculty member had just parked three spaces way and was getting out of her car when she saw Hank fall. She ran over, took one look at his ashen face and dialed 911 on her cell phone. Two minutes later she heard a siren begin to wail, and when a campus fire truck pulled into the garage she backed away and simply pointed to the space between cars where Hank lay. The EMT team detected a weak, erratic pulse, so they put him on a respirator and rushed Hank to the ER of the Bloomington hospital without much hope but with the respirator helping his lungs pump air. With no outward symptoms, the supervising resident ordered an EKG for the heart and an EEG for the brain. The heart beat was weak, with an arrhythmia characteristic of a cardiovascular infarct. The EEG picked up only the faintest brain activity.

Michelle took a call from the hospital minutes after the ambulance arrived. On her way to the hospital, she left Avalon in Bota's care, a detour of only 2 or 3 min. Fighting traffic that seemed to intentionally block passing lanes and stop too soon and start too slowly at lights, she thought, *It's a mistake. It's not*

Hank. But the man lying peacefully in the intensive care bed was Hank, or at least his body. Almost immediately she thought, *Avalon can't lose another parent.* She bent down and whispered urgently in Hank's ear, "*Avalon— Avalon needs a father.*"

The next few days passed in numb duties, with trips back and forth to the hospital while she slowly came to grips with the realization that anything recognizable as Hank had left his still living body. She met with the physician who had first seen Hank when he was rolled into the ER, a young Indian man with his name on the ID fastened to his white physician's jacket: Panganamala Venkatarao, MD. He smiled when he saw Michelle trying to puzzle out his name. "Professor Murphy? You can just call me Rao."

Michelle smiled back. "Thank you, that helps. I didn't hear an accent?"

"My parents moved here from Mumbai, so I grew up in Bloomington."

Michelle nodded. "How did you know who I was?"

Rao's smile broadened. "Believe it or not, five years ago in my senior year I was a student in your astrobiology course. It was unforgettable."

Michelle sighed and shook her head. "It's a small world, isn't it. But now I have something to learn from you. Do you have any idea what caused Hank to collapse like that?"

"His EEG was basically a flat line, as expected, but to determine whether a stroke or tumor had caused the collapse, I also ordered an MRI. There were no obvious tumors, but the results were very puzzling. We are used to seeing single embolisms, clots that block blood flow to the lower leg, for instance, or to the lungs. Some heart attacks occur when a clot detaches from elsewhere in the body and reaches a coronary artery. But your husband had dozens of embolisms, any one of which could have been fatal. This is so unusual that I'm going to write a brief report for JAMA."

A week later, Michelle gave Hank's living trust document to the legal representative at the hospital who read it carefully. Then, as next of kin, she signed where required to remove life support. As his body shut down and passed from life to death, she held his hand the way she had sometimes held Avalon's hand as the child went to sleep. She would remember for the rest of her life the cooling of his hand.

In the weeks that followed, Michelle slowly came to terms with the fact that the human being she loved was now inexplicably dead. If dying is so easy, even for someone as vital as Hank, she couldn't help asking herself why she was alive and Hank wasn't. Somehow the answer seemed hiding in their adopted daughter, her unusual intelligence and boundless enthusiasm for living. At times Avalon asked, "Mom, why are you staring at me?" Michelle said only, "Because I love you," and that was, in a way, the true answer. Like nothing else Hank and Michelle had done, Avalon had been their joint choice without reservation, their gift to each other. What Michelle had begun to regret most was an unsettled argument that had frequently interrupted the harmony of her life with Hank the year before his death. The subject was Avalon.

Chapter 7

In 2008, when they had both been new PhD's at the University of Indiana, Hank and Michelle had decided to adopt a child. Like many Americans, they wanted an adopted infant who would know only them as parents. America's limited choices and unlimited paperwork and the alternatives suggested by friends had convinced them to look abroad.

The university had a very strong program in Central Asian studies, and a particularly charming young anthropology graduate student named Botagaz from Kazakhstan had pleaded with them to adopt a child from her country. "You cannot imagine how sad the kids are in our children's homes," she said. "And when they leave, they have nothing. I can show you studies. You know the profession chosen by half of the girls—prostitution. There is nothing for them." In her dark brown eyes tears began to gather and her delicate lips trembled. She laughed bitterly and said as if in a dream, "You know some people say we are the descendants of the great Genghis Khan." Suddenly she focused intently on Michelle and Hank, took a deep breath and said, "You will make a huge difference in this child's life." Botagaz's voice always had a musical quality to it, and now her voice, her eyes, her lips, her hand on her heart made an irresistible harmony.

Michelle asked the question that decided matters. "Botagaz, did you grow up in a children's home?"

The corners of Botagaz's mouth tightened, and she suddenly slumped into Michelle's open arms. Her body heaved with sobs she had never permitted before. Although Botagaz was a grown woman, her head on Michelle's shoulder, the warm tears from within, the futility of words and the comfort that flowed out of her own arms transformed the two grown women for an intense moment into mother and child. In that moment a silent promise became a binding fact of Michelle's future. Holding Botagaz tightly, Michelle knew that she and Hank would have a Kazakh baby girl.

Two months later, with the necessary paperwork and visa in hand, Michelle and Hank had flown halfway around the world to Almaty, a city of one and a half million people that had been founded a hundred and fifty years earlier as the Russian fort of Vyerni on the flat steppe of the southern Siberian plains. Under the Soviet Union the settlement had sprawled its way up the gentle slopes of a vast outwash plain that flanked the great white wall of the Celestial

Mountains. Those mountains known as the Tien Shan separated Kazakhstan from Kyrgyzstan to the south and beyond that Tajikistan and Afghanistan.

At the airport a translator and a driver waited just outside the customs door, the driver holding a sign with their name on it. He was a man in his fifties, wearing a plain blue shirt and black pants, his face as brown as wood and as hard as stone. The translator smiled broadly and called to them in pleasantly bouncy English, "Welcome to Almaty. I am Zaure and our driver is Kairat." She said *our driver* in a way that clearly meant that she, Michelle and Hank were masters and Kairat their servant. Kairat, in turn, soon proved that he considered himself the owner of any street he drove on. He honked at other cars, swore out the window at their drivers, and swerved toward jay walking pedestrians as if they were targets. Michelle said to Zaure, "Could you ask him to calm down and go a little slower?"

Zaure's answer was, "He has been a fireman and a policeman." Michelle frowned and thought to herself, *as if that should settle the matter.* Understanding was lost in a cultural crevasse. In the translator's world policemen and firemen were dangerous, often violent. "Besides," Zaure said, "he is a country Kazakh and these people drive the way they ride horses."

The comment brought back to Michelle's mind Botagoz's bitter comment about Kazakhs being descendants of Genghis Khan.

In the morning, after a short sleep in a rented apartment, they found Children's Home 45 as alien as Kairat. They knew when they walked down the first corridor that a child's life there was as sour as the smell of cooked cabbage that permeated the air. What Botagaz had not prepared them for was the corruption whose practitioners turned out to be almost anyone providing a service or signing a necessary document. Like hungry but quiet little birds or smirking bullies, they waited to be fed. On the third day, out of earshot of their translator, Michelle and Hank talked about going home and starting over, but then they met a 1-year-old girl whom they could not abandon.

Or was it that the infant would not abandon them? Michelle understood in her first hour with the infant that this tiny child was no ordinary 1 year old. Michelle was sure the girl somehow knew who she was, what she was thinking. She told Hank as she put the infant down in her crib. "I have the feeling that somehow she is still holding on to me."

Hank said, "It's what children do with mothers—bond."

"But I'm not her mother," Michelle, the scientist, observed.

"But she wants you to be."

"I really think she does," Michelle said. "But watch her eyes. They hardly move. She's staring right into me. I get the feeling she is trying to say something."

"Is that bad?" Hank asked.

"No, but there's a strange intensity about her." They had been warned that foreign orphanages often tried to get rid of children with mental or physical problems.

The most interesting thing about this child was that when the matron picked her up to return her to the nursery, she made baby noises but noises that sounded like something more than babble. Nothing they could understand, of course. When they asked their translator Zaure about it, the answer was, "Food, they always talk about food."

"What is her name?" Michelle asked.

"I can tell you only her first name," the matron said. She spelled out a long incomprehensible word.

Michelle asked her to repeat it, and Zaure said she would write it out for them in English. Michelle looked at the word and said, "Can we call her Bota for now?" She knew from Botagaz that this was an affectionate name that translated to 'little camel'.

The matron chuckled. "Call her that if you like. We have lots of Botas."

As they played with the baby, both Michelle and Hank had the impression that Bota was trying to talk to them, though that would verge on the miraculous. They laughed at themselves and joked about having sentimental hallucinations, but they decided to finish the process so Bota could come home with them.

The adoption advisor they had consulted told them both to wear a money belt stuffed with fresh hundred dollar bills, and now they understood why. They paid exorbitant rent for a decrepit one-bedroom apartment on the fifth floor of a dingy concrete building whose elevator was the size of a broom closet. They paid for the translator who translated only half of what was said. They paid the driver Kairat for his dangerous driving. They paid several notaries who had to examine and approve paperwork. And finally they paid for a "donation" to the orphanage for health exams, for their own background check which took no more than an hour after they paid. Each time they hesitated, Zaure would say, "It is necessary. It is our way." And Michelle and Hank had understood from the first day that only by collaborating with the corruption could they leave Kazakhstan with the baby.

They paid, and they signed many papers they never saw again. Some papers they were told they would need, suddenly became unnecessary after Zaure disappeared into a certain office with a new hundred dollar bill. Zaure surprised them late one afternoon when they thought they were done for the day. After answering a call on her cell phone, speaking in Kazakh, she told them, "I must take you for a background interview."

The interview was in a government office building with a pharmacy and clinic on the ground floor, an elevator that smelled strangely antiseptic, and unusually nice carpet on the third floor where Zaure guided them down a corridor. Their interviewer was a middle aged, slim Kazakh man in a very nicely tailored gray suit. He stood up and came around his desk to shake their hands and introduced himself as Dr. Serik Faizilov. Instead of sitting behind his desk, he buzzed an attendant to bring tea and sat with them in upholstered chairs around a low table.

Afterward both Michelle and Hank agreed this had been the most pleasant encounter with officialdom and a strange interview, more like a job interview about their academic backgrounds, achievements, and research interests. Dr. Faizilov said it was not unusual to have a professional with similar background as future parents verify their credentials—a businessman for business persons, a scientist for scientists. When the interview was over, they waited for the usual request for a "donation" and found it strange that the good doctor simply wished them a pleasant journey home.

On the morning they prepared to sign the final papers—well, the third or fourth set of final papers—the ones after which Zaure assured them no more obstacles would arise and no more payments would be demanded—Hank and Michelle sat in the apartment drinking coffee, each thinking. Michelle said, "I really want this child, I really do, but. . ."

Hank said, "But we have to ask some hard questions."

"Yes. And before we see Bota again. Or before she sees us maybe I should say."

That night they awoke, startled because the bed was shaking and a strange deep rumble penetrated the room from outside. They jumped out of bed and stood in the center of the room, Hank with his arm around Michelle. After a moment's confusion, Hank said, "Earthquake. Hope it's not a big one." But the shaking died away almost as soon as he finished speaking. They considered going out into the street where it might be safer if there were aftershocks. Michelle walked to the window and looked down, but the streets were empty.

Michelle said, "I wouldn't want to be in these buildings when they have one like they had in 1910."

Hank nodded. "It would be like inside a trash compactor."

The next day no one mentioned the earthquake. In the notary's office where they sat before a small pile of papers, accompanied by an adoption agency official, Michelle asked the questions. "Did the baby have any health problems?"

"Of course not," the official said. "Our homes provide excellent health care and every child is fully examined when taken in and any time our staff notices something questionable. Every child has had the usual vaccinations."

Michelle asked, "Do you know who the birth parents are and if she has any brothers or sisters?"

"These things I don't know, and they are not public records. But you can be assured that she would not be in the Children's Home and available for adoption if there were any uncertainty. Do you want to adopt another baby? That would require . . ."

"Have you done any developmental testing?"

The official raised his eyebrows in question.

"To see if she is developing normally—compared to her peers."

The official shrugged. "Our staff sees these children every day. They would notice something not normal. We do not deceive people."

Michelle and Hank exchanged a glance and a small smile, knowing that he was deceiving them.

They asked a few more questions, then began to sign papers, insisting on a full translation despite an obviously impatient official who rolled his eyes each time they asked for another translation.

They took Botagaz's advice and paid with their last hundred dollar bill for VIP treatment at the airport departure lounge. It was worth it. The fee let them detour around the various gatekeepers where bribes were often demanded. The ticket inspector weighed their bags but didn't look at the scale. The security inspector glanced at the X ray screen but waved them through, and the pleasant uniformed passport inspector gave their papers only a cursory look, smiled at the baby, and wished them the customary "soft landings".

During the stop in Amsterdam Michelle took the baby into a restroom that had a diaper changing table. As she dusted the baby's bottom she said, "There you are—feel better?" With those words she discovered that they had not adopted a physically normal baby. As Michelle looked up to see if the baby was smiling, the baby looked back—only with the right eye. The left eye was looking in another direction, toward the container now holding a soiled diaper. Michelle knew about "lazy eyes," in which one eye does not move with the other. But the baby's right eye was certainly not lazy. It moved, stopped, moved again. And again. Then both eyes focused on Michelle and the baby rewarded her with a lovely smile.

Michelle warmed to the smile and to her success in this small act of motherhood. She put her own smiling face an inch from the baby's and said, "You're going to get a full check-up when we get home, and we'll see about fixing that eye."

On the flight to New York Michelle sang the baby to sleep with a song her grandmother had sung for her.

Bye, baby Bunting,
Father's gone a-hunting,
Mother's gone a-milking,

Sister's gone a-silking,
Brother's gone to buy a skin
To wrap the baby Bunting in

The baby fell asleep. When she woke up a few hours later she looked at Michelle and made sounds something like cooing, but Michelle was sure she heard something of the rhythm of the lullaby. She asked Hank to listen. She rocked the baby in her arms and sang part of the lullaby. The baby nodded its head and made sounds that clearly repeated the beat of the lullaby.

"Musicophilia," Hank said. It was the title of a book by psychiatrist Oliver Sacks about how music works in the brain. "Maybe she will be the youngest member of Yo Yo Ma's Silk Road Ensemble."

Later in the week the Bloomington pediatrician said they not only had a healthy if slightly underweight baby, but that the eye movements, while rare, did occur. Since she could focus both eyes on one subject when she wanted to, he wouldn't worry. "We are learning more and more about the variety of ways the brain's wiring develops. Don't consider this a flaw or a problem unless it begins to cause trouble."

Hank asked, half seriously, "Do you think it will be an advantage if she learns to play piano?"

The girl whose name they changed to Avalon began to say a few English words only a month after they had returned to Indiana. And she did develop her own version of Bye Baby Bunting. One day when Hank had gone to the university Avalon said to Michelle, "Daddy's gone a hunting." By two she was speaking whole sentences and talking about everything she saw, though she still pronounced English like a 2 year old. By three her pronunciation was good. Hank and Michelle even joked with Botagaz that Avalon would have a musical voice like hers. Avalon called Botagaz, Apayishka—the Kazakh word for Aunty. This amazed Botagaz, because it meant that Avalon recalled some of the language she had heard just for a few months in her first year of life.

When Hank or Michelle read to Avalon, she looked not only at the pictures but at the words. One night after turning out her light, Hank gave Michelle a big kiss and said, "Avalon can read." He had taught her "cat" and she had learned hat and bat and rat quickly. The next night as Hank stood up to leave her room, she said, "The cat ate the rat then slept all night in a hat. What do you think of that?"

Avalon soon fell in love with books, and reading became a passion. She would grab any book in sight and study it intently page after page, looking from picture to words and back many times. They found that she easily learned to find the words for the images in a picture once they showed her the connection—not only one syllable words, but two and soon any word.

Avalon began carrying one or two books with her everywhere. She insisted on having books by her bed. She read fast, super fast for a child. They discovered she seemed to read adjoining pages simultaneously, one with each eye. And when she read, she seemed to enter a trance—hearing no one, seeing nothing, physically present but mentally gone into the world of the story. But when she finished, she came out talking non-stop.

Michelle was the first to worry that the intensity with which Avalon applied her intelligence might be a form of autism. Maybe she was a reading savant. A colleague studying the genetic basis of autism suggested that Michelle test Avalon for any of the known genetic markers. Work by Rita Cantor at UCLA had found several genetic variations typical of autism. The test on Avalon showed none of these markers, but an anomaly that might be yet another marker of autism. That test and its analysis resulted in Michelle's first publication in genetics as co-author with her colleague. The two scientists, using a small grant from the National Institute of Health, compared Avalon's genes with those of 50 autistic children and 50 normal children. None of the autistic children had Avalon's unusual marker. That was a great relief to both Michelle and Hank. However, neither did any of the normal children, including a couple of Kazakh children of graduate students and a visiting professor.

Michelle and her colleague published their findings in *Nature Genetics*, "Unusual Marker Sequence for Autism Spectrum." What Michelle and her colleague did not include in the paper was the nature of the translation product, perhaps a transcription factor, located on an open reading frame of Chromosome 7. Michelle decided to call that protein Avalin. That's when Hank and Michelle began the argument about Avalon's future.

Hank was all but certain that the unusual marker in Avalon's genes endowed her with her unusual intelligence. Michelle said maybe. Hank wanted to do more testing. Michelle said, "I'd like us to be mother and daughter, father and daughter, not scientists and subject." Hank argued they could be both. Michelle said science could wait. Hank wanted to go back to Kazakhstan with Avalon and find her parents and her brothers and sisters. If she had brothers and sisters.

"I don't want to share her," Michelle said, "not with people who abandoned her."

"We don't know that," Hank said.

"And I don't want to find out," Michelle parried, "even if they would tell us, and they won't."

"That is probably a matter of finding who we have to pay and how much."

Michelle agreed, but shook her head. "Taking her back and finding her family—that might be good for science—even I'm curious. But it would not be good for Avalon. Maybe when she grows up, she can decide for herself. I don't want to turn her into a scientific specimen or worse, a freak."

Hank replied that the longer they waited, the less likely they were to find anyone in the family. Michelle accused Hank of putting science before the welfare of their new daughter. Hank accused Michelle of being afraid of the truth. They argued their two positions back and forth for months, sometimes into angry silences, although those silences never again lasted through a night.

Suddenly Hank lay unconscious in the Bloomington hospital. As Michelle attended him, the argument hung suspended in her mind. Before she watched his body die, she had promised herself that one day she would return to Kazakhstan with Avalon and look for members of her family.

Hank had been so much a part of their family life in Indiana that after she emerged from the first weeks of shock and grief, she felt his presence everywhere. Sometimes she dreamed that she and Avalon were in a plane flying to Kazakhstan, with Hank sitting across the aisle, happy that he had finally won the argument. With the right subject and a drink or two, she often felt Hank at her side, with friends talking to Hank as much as to her. Except he could not reply. She spoke for both of them, sometimes giving his view, then hers. His ghost made his absence acutely painful. One evening as a powerful storm cell hurled golf-ball sized hail stones against her roof, she looked in on Avalon, sleeping soundly in her bed. Upstairs the thunder of the ice blocked all other sound. Suddenly she became aware that Hank's presence had disappeared. She looked down at Avalon and said, "It's you and me now, and it's time to go." She decided with the relief of finality to look for a position far from Indiana. The hail storm suddenly stopped and she smiled.

Michelle would present herself from here on as a geneticist. With Botagaz seeming to be happy staying with Avalon, Michelle interviewed at the University of Texas, at Duke University in North Carolina, at the University of Florida, at Washington University and with some reluctance at the University of Oregon in Eugene where rain fell almost every day for 7 months between October and May and where the department seemed very uneven. To the surprise of her Indiana colleagues Michelle turned down a very rich proposal from Duke and accepted Oregon's less lucrative offer.

While the Oregon biologists wanted Michelle to fill a hole in the curriculum, what first pulled her toward the position they offered was the warmth and integrity of the man who had recruited her—Donald Koskin. In his mid 40s, he was lanky, loose, confident, and soft spoken. She believed his promise that she would have freedom to pursue her own research path. Don's wife Jodie, who hosted a private dinner for the three of them, sealed the deal. The obvious bit of salesmanship was an all-Oregon dinner beginning with an Oregon champagne, salad from a local farmer's market, Oregon artichokes with

Oregon Salmon. When Don announced the dessert would be a pie he baked with Oregon's sweet Marion berries accompanied by Oregon coffee, Michelle laughed and told him that, "Somehow I think I get the point of the menu."

When Don had taken dishes to the kitchen and was preparing the dessert and coffee, Jodie lowered her voice. "Michelle, the point I want to make is that Don and the department need you here. We know that Oregon's salaries are less than you could make elsewhere, and the faculty are still struggling to make it a top tier biology department, but you really do fit so well. You are their missing link."

Michelle wondered how she could like these two people so much after so little time, even feeling some obligation to them. A second reason for accepting the position was one she never told her colleagues and friends. At the dinner with Don and Jodie she had found herself talking about Hank and about Avalon. They had listened with genuine interest. Michelle showed them her favorite picture of Avalon, Hank and herself. Hank and Michelle leaned together shoulder to shoulder grinning into the camera while and Avalon sat high on Hank's shoulders looking upward to something out of sight. Michelle said, "Whenever I look at her playing or reading or sleeping I ask, "Who are you?" Jodie said, "That will take some time, and someday she will answer your question. What I know for sure is that you can help our students find out who they are."

Chapter 8

2020 January, Eugene Oregon

Michelle's signature on the CIA confidentiality form made her feel as if she had drunk a magic potion, not a good one. She woke up thinking about it. At work it churned in her mind, demanding attention, but when she stopped to think about it, she came to no conclusions and it was still churning 3 days after Evan and Rhonda had visited. She decided to seek advice.

"There is a name for this kind of day," Don Koskin said to Michelle several days later as she sat watching him drink his third cup of morning espresso. "Everything starts off as usual—I get up when it's dark. The rain is falling. The morning begins like a cold wet blanket. It's a nothing day, then suddenly it's a something day and out of nowhere things come together."

"Serendipity," Michelle offered. She wore jeans and a black wool sweater, and sat leaning forward, one elbow on top of her crossed legs.

Michelle said. "I have something to show you, and I need your advice."

Outside wind beat the rain against Koskin's window in slow, powerful gusts, like the exhalations of an angry sky. An hour and a half earlier Michelle had hailed him as he was coming in the building and asked for a few minutes to talk. When they had hung their coats in his office, she drew an envelope from her bag and extracted her invitation from The Kazakh Institute for Biology of the Future Named "Asfendiayarov". She handed the invitation to Don who sat down at his desk and read it. He began to smile.

"Wonderful," he said. "Congratulations. You'll go of course? With Avalon?"

"I think I will," Michelle said. "And with Avalon. Hank would be happy, and I think she's old enough."

"Why wouldn't you go?"

"You know when Hank and I were there to adopt Avalon, we didn't exactly leave thinking it was a bastion of civilization, though I guess a lot has changed, Borat notwithstanding."

"Oh, yes, Borat," Don mused, recalling the movie of the shock-jock, faux Kazakh journalist reporting on America with the pretense of innocent bawdiness.

"And don't you think the invitation is a little odd," Michelle said. "Usually you know the people who give out invitations. Anyway, this one is directly from the head of the Institute—Dr. Arman Akenov."

"Rings a bell," Don said, and paused to think. "It's ringing louder, coming through. . . Ah yes, 'the white knight' of Kazakh science." That's what he was called in an article I saw—*Time* maybe. I've read two or three papers by Akenov because of my work on early domestication of horses. He's very big on horse breeding. He says the domestication of the horse was the first way in which Kazakhs changed the course of history."

"Hm. I suppose it did," Michelle agreed, "but did the Kazakhs do it? Where did they come from?"

"Their ancestors are Turkic and Mongol people, but exactly when and how they became distinctly Kazakhs is debatable. Four, five, maybe even six centuries ago."

"So they didn't domesticate the horse?"

"Who knows? We can't say that the Kazakhs or their direct ancestors weren't the domesticators, it's just that no one can prove they were. The horse was tamed about 5,000 years ago, and very few people have been in the same region for more than a millennia or two."

"What about your Denisovans?" Michelle asked.

"You'll have to read my paper," Don said. "And even then you won't know, but I can guarantee you they were in Kazakhstan before the Kazakhs."

Don had come in to work that morning looking forward to finishing a review he was writing for *Nature* about the emerging genome of the Denisovans, a third human species or perhaps a Neanderthal cousin that existed along with humans 50,000 years ago. The initial evidence had come from a finger bone discovered in Denisova Cave in Altai Krai, a region of south central Russia that extended into Kazakhstan. Other bone fragments had more recently been discovered in another cave in Kazakhstan.

"Denisovans," Don mused. "One more humanoid that came and went. How many others do you think? Hey, did I tell you the results I got from my 23AndMe DNA analysis?" He didn't wait for an answer. "I'm 3.5 % Neanderthal. That's considerably above average. Would they be proud of me?"

Michelle smiled indulgently. "So that's what Jodie sees in you. That's why she found you in a cave." Don and Jodie had in fact met in a cave. People sometimes joked that they looked like twins—both tall, blond, lanky, with the slightly loose jointed motions that often go with height. They were adventurous, and had met while exploring a remote passage of Mammoth Cave in Kentucky. Koskin had accidentally caught his hard hat on a low hanging stalactite that broke his lamp. Jodie let him borrow her backup, a flashlight, and she led him on a two hour crawl back to the entrance. Years later, when asked how they had first met, Koskin would answer, "When I met her, I saw the light."

"I've looked him up—Akenov," Michelle said. "The Institute and his foundation too. All part of a national plan for improving resources using modern genetics and genomics. Some interesting work on domestic animals and plants, and lots of papers to his name. But he's made some pretty far out statements, such as Kazakhs domesticating the horse and inventing the chariot, but also that Genghis Khan was a Kazakh, not a Mongol."

"He's not the only one," Don said. "I guess every country wants a glorious and long history, especially if other people ruled if for a few centuries, the last one being the Soviet Union full of Russians."

"Is it credible—about Genghis Khan?"

"Since nobody has found the body we have his DNA only from descendants. A lot of them could be today's Kazakhs."

"Well, Genghis Khan might be proud of Akenov too. On the web site all kinds of praise from other scientists, members of parliament, and the president—their president—says Dr. Akenov is world famous."

"Maybe world famous in Kazakhstan," Don said. "Small pond, big fish."

"Big money too. Besides the honorarium there's first class air fare including Avalon's. They know she is Kazakh. Isn't that a little strange?"

"You researched them. They researched you. You are all over the Internet, you know. Not to mention the faculty pages. And I'm sure their government has computerized records easy enough to search if you have access."

"All that, sure. And remember when I ran that little sequence on Avalon when Hank and I thought she was so intense that maybe we were seeing some strange kind of autism?"

"I even recall the title: "Unusual Marker Sequences in Autism Spectrum" or something like that. Published in *Nature Genetics*, right? And it wasn't just Avalon. Didn't you include fifty or so local autistic children and a control group?"

"Right. I didn't use her name but I did say that the marker was in a child of Central Asian descent. That's when I started to study evolutionary genetics—I wanted to understand more about intelligence—not just IQ, but other dimensions. Sorry, I think we got off topic."

"Now I remember. You were using Rita Cantor's work at UCLA and the copy number variations she found in a lot of autistic kids."

"Damn, your memory is good."

Don smiled, looking past Michelle, beyond the window into the branches bending and unbending in the rain storm, more wind than rain, as usual. "My Neanderthal ancestry," he said dryly.

"I think I was more relieved than Hank was when the CNVs didn't show up," Michelle continued.

"But you found avalin . . ."

"I found a protein I wanted to call avalin—you're right. But I found it only in Avalon."

"Chromosome 7, open reading frame. I still think you should publish something on that."

"If the literature showed any other human being having it or it being linked to some problem, I would. But I don't think Avalon's ready to be a biological specimen in public yet."

"Motherhood before science," Don said with admiration. "But about the invitation for you and Avalin, I mean Avalon. . . sorry."

Michelle laughed, but she was serious when she said, "Even Avalon doesn't know about avalin. Not yet. I'll find out more about it one day."

"Both Avalon and avalin are safe with me," Don said. "Now, back to Kazakhstan and the invitation."

As they talked, Michelle wanted to tell him about Grable and Lipkovich and their interest in Akenov. What she really wanted to say is, *I am a spy.* What harm would it do? Don could be counted on to say nothing. She had signed a contract, however, and while she didn't think she would die for secrecy, neither would she break a promise for the comfort of company.

"I just thought of something that should interest you," Don said.

Michelle just raised her eyebrows and gave him a go-ahead-and-tell look.

"The department has a graduate student from Kazakhstan, and as I recall, he was recommended by Dr. Akenov. Have you ever met Bakhyt?"

"Not that I recall, but maybe I should."

"I know what he's working on—genetic connections between Central Asians and the languages they speak. Suppose I see if Bakhyt is around and what he knows?

"Do it," Michelle said.

Koskin reached for his phone, punched a few numbers, and after a brief conversation, nodded at Michelle. "Bakhyt was in the lab. He's on his way up. While we're waiting, would you like an espresso?"

Michelle shook her head. "Never touch the stuff. But go ahead and I'll watch."

A few minutes later Koskin's device hissed and steamed out a small cup of the caffeine-laden espresso he liked. While he was carefully carrying it back to his desk there was a knock on his door. "C'mon in Bakhyt!"

Bakhyt slipped in, carefully closing the door behind him. He was short and trim, his black hair cropped close to his head, a thin mustache across the top of his mouth. From his pressed white shirt to his well shined black shoes, everything about him suggested a passion for order and attention to detail. Michelle immediately noticed the same dark brown eyes as Avalon's under the same thick black eyebrows. Central Asian eyes, she had observed long ago, were wider and not as oriental as Chinese and Japanese eyes.

With a noticeable accent Bakhyt asked, "You wanted to see me Professor Koskin?"

"Yes, thanks for coming up. Bakhyt, I want you to meet Professor Michelle Murphy."

Bakhyt put his right hand over his heart and bowed slightly to Michelle, then as soon as he saw her outstretched hand, he smiled with slight embarrassment and shook her hand.

Koskin went on, "Michelle is traveling to Kazakhstan next month, to Astana in fact, and wanted a little background information. Go ahead and sit down over there, let's see if you can help her. And how about an espresso?"

Bakhyt again put his hand on his heart and bowed slightly as he said, "Oh, yes, I very thank you."

For the next few minutes, Michelle asked Bakhyt about his research , then gave him an overview of her own work. She asked where he had grown up, and he said, "Keskelen—lots of Turkish peoples there—you know Kazakhs and the Turkish we are cousins? Turks came from Mongolia."

"I remember lots of Turkish companies in Kazakhstan, and I know Kazakh is a Turkic language."

"You visited my country?" Bakhyt asked with obvious pleasure.

"We adopted my daughter there—almost thirteen years ago, in Almaty," Michelle explained.

"Our loss it is your gain," Bakhyt said.

Michelle thought she heard a slight bitterness despite the smile.

"It's a global village now, isn't it?" she proposed. "So I hope we'll all gain. And we're both invited, my daughter Avalon and me, to visit Kazakhstan. That's what I wanted to talk to you about."

The espresso seemed to warm Bakhyt. "Oh, yes? Please."

"Do you know The Institute for Biology of the Future?"

"Yes, of course. A special organization in our country."

"Then you must know of the director—Dr. Akenov, Arman Akenov."

"Yes. Of course, every scientist in Kazakhstan knows who he is. He is world famous."

Don caught Michelle's eye, and she raised her eyebrow very slightly. "I've seen his papers listed in a lot of bibliographies and citations. Very impressive. I don't see how he could do all that work in one lifetime."

Bakhyt didn't say anything for a few seconds. He sipped his coffee and looked out the window. "I think he works very long. For him is a very big laboratory."

Koskin said, "Government funding his research?"

"Oh, yes, of course. Kazakhstan you know is very rich in oil and gas. And minerals. It is said we have every metal on the periodic chart of elements. Gold too. We even have Golden Man story all children learn."

"And the Good Shepherd Foundation?" Michelle asked.

"Yes. It is his foundation for improving our health system. Especially for children."

"Did you ever meet him?"

Bakhyt looked away, as though he did not want to answer this question. Then he replied, "Yes. He was interviewer when I applied for Bolashak scholarship."

Don followed up, "So he is partly responsible for your being in the U.S.?"

"I say so," Bakhyt said.

Michelle leaned forward and asked almost flirtatiously, "Bakhyt, what aren't you telling us about Dr. Akenov?"

Bakhyt hesitated noticeably before he said, "Just what everybody knows."

"But not me," Michelle said. "What would you say everybody knows that I should know?"

This time Bakhyt's eyes studied the ceiling as he thought for a while. "I think the late President Nazarbayev thought very highly of Dr. Akenov. I think our new president will too."

"And you?" Michelle asked largely because the omission of his own opinion was an obvious evasion. Or was it just a cultural difference, the kind of question that in Bakhyt's world required modesty and deferral.

"President is very powerful man. So Dr. Akenov is also powerful man."

Bakhyt had finished his coffee and put the cup on the table with the espresso machine. Both Koskin and Michelle could see he was uncomfortable and ready to go.

Michelle said, "Different subject—will my daughter—she's 13—enjoy Kazakhstan?"

Bakhyt didn't hesitate before saying, "Yes. She will very enjoy country of her brothers and sisters." He caught a slight wrinkle between Michelle's eyes and explained, "Brothers and sisters, as you say in nationality. I am one."

"I understand," Michelle said. "Sorry, but maybe one more question."

Bakhyt waited.

"Would you have any time to give us a crash course in Kazakh? We should know some basic phrases at least to be polite."

Now Bakhyt smiled very broadly. "It would be my great pleasure," he said. Michelle promised to call him.

When he had left, Koskin said, "Sounds like the two of you have decided to go."

Chapter 9

2020, Eugene Oregon

In early January, the first week of the quarter, Professor Don Koskin had settled into the "warm up" routine for his afternoon graduate course in human evolution—a brisk 20 min power walk from his house to the campus Starbucks. The counter staff knew what to pour for him, and he sat at a table near a wall so he could look over the room. Among the dozens of students talking, flirting, studying, and working their laptops, tablets and cell phones Koskin browsed for identifiable primate behavior. He usually left with at least one or two observations with which he could warm up his lecture on human evolution. He could state the prime message of his course in a sentence: civilization is a very elaborate society of formerly wild animals who had tamed themselves. He never did say it. A student could pass the course only by demonstrating that he or she had discovered this truth in some way.

He watched a coed picking lint off a boy's shirt. A famous Oregon Ducks basketball player loomed in the doorway, wearing his garish yellow and green jacket. He paused, surveying the room, grading the females, assuring himself he was the biggest male. A serious guy involved in serious business. Evolution is very slow, Don reminded himself, and made a note to develop that into a lecture about the slow and irregular pace of primate evolution. That might not last, he mused. Around the world hundreds, maybe thousands of scientists were discovering new ways to modify genes and the plants and animals that possessed them. At Harvard George Church's lab was considering ways to recreate the wooly mammoth.

Don checked his watch—3 minutes behind schedule. He took a last sip of his latte, grabbed his briefcase and headed out the door at a brisk pace, enjoying the stretch and the crisp winter air.

When undergraduate students at the university spoke about Don Koskin, or one of the courses he taught, they would say something like, "You know, the good professor." Students appreciated him not just for rigorous and entertaining lectures, but for an unusual degree of professional generosity. His grading was tough but fair, and at the beginning of each course he announced the rules of his unique process by which students could change a grade on one of their tests. Anyone could appeal a grade, but the entire test would be graded again, so that opening an appeal might lead to a grade reduction. Although an appeal rarely did lead to a reduction, it had happened, and those few cases were passed around as legend. Don was glad he had made those decisions. He seldom had a frivolous or arbitrary appeal.

The administration had once asked him to consider taking the post of University Ombudsman. He had declined. Koskin guarded his precious time in the laboratory, and also made time for his wife Jodie and for their friends. His open and untroubled love for Jodie was the envy and marvel of his peers.

When he quietly opened the door at the back of the small, windowless classroom, he paused. His students had their laptops open or tablets turned on, mostly looking at FaceBook pages, email, or playing games. One young woman had the class web site on her screen. As he walked up the aisle to the desk and whiteboard he saw the class website popping onto more screens. Fifteen years ago with a new Ph.D. in hand he had been lecturing to students who bent their heads over paper notebooks splayed on their desks, pens or pencils in their hands. His own visual aids had been what he could write and diagram on the whiteboard or plunk down on the overhead projector. He welcomed electronic technology and he was good at it, but he occasionally wondered whether today's students were any better off by having all the class material fed to them with electronic spoons. He saw his job as giving them mild intellectual indigestion, while simultaneously hinting where they could find the mental Alka Seltzer.

The projector was still on from the previous class, so he plugged the cable into the video port of his iPad. While the projector and iPad shook hands and agreed how to talk to each other, he looked over his class. From the roster, he knew there should be seven women and two men, but he saw Michelle Murphy sitting in the back of the room. He nodded at her, happy that she had accepted his invitation to sit in on this particular class. Her black hair was tied in a simple pony tail, revealing small hoop earrings. Michelle smiled and with a brief twist of her hand that she was ready.

"Folks, we are lucky today to have a distinguished visitor. This is Professor Michelle Murphy from the Department of Genetics. She specializes in human genetics but also does super secret stuff over at TransTek. Come on up Michelle. I'll bet you have something interesting to tell us about FOXP2."

Michelle stood and made her way to the front of the room with her briefcase while Koskin pressed the advance button. A new slide appeared on the screen.

Molecular evolution of FOXP2, a gene involved in speech and language. Enard W, Przeworski M, Fisher SE, Lai CS, Wiebe V, Kitano T, Monaco AP, Pääbo S. Nature 418: 869–72 (2002).

"A functional genetic link between distinct developmental language disorders." Vernes SC, Newbury DF, Abrahams BS, Winchester L, Nicod J, Groszer M, Alarcón M, Oliver PL, Davies KE, Geschwind DH, Monaco AP, Fisher SE N. Engl. J. Med. 359: 2337–45 (2008).

Michelle paused for a moment, looking out at the nine students, waiting for their full attention. "The story begins in Germany, in 2001 when I was visiting Svant Pääbo's lab to give a talk. By the way, this is the same group that sequenced the Neanderthal genome. Do you remember that? They made international news when they announced that Europeans and Asians have a few percent of Neanderthal genes in our genomes, most likely from inter-breeding some 50,000 years ago. Professor Koskin, by the way, told me that he is 3.5 % Neanderthal, of which he seems to be surprisingly proud."

This got a loud laugh, and Koskin turned and waved at the class.

"Anyway, they had just begun to work out the amino acid sequence of the human and chimp FOXP2 protein, and told me in confidence that two of the amino acids distinguished chimps from humans. In their 2002 paper, they speculated that it might be related to language. This was absolutely nailed in the 2008 paper by Vernes et al. Does anyone know that story?"

Polly raised her hand. "I recall reading a couple of years ago about a strange genetic thing that makes people unable to talk."

"Correct. Good memory! That's from the Vernes paper you see up there, and I recommend that everyone read it. Don, what do you have on your next slide?"

Koskin laughed while he pressed the advance button. "I must have been reading your mind. Here's the abstract of the Vernes paper."

Rare mutations affecting the FOXP2 transcription factor cause a mono-genic speech and language disorder. We hypothesized that neural pathways downstream of FOXP2 influence more common phenotypes, such as specific

language impairment. We tested for associations between single-nucleotide polymorphisms (SNPs) in this gene and language deficits in a well-characterized set of 184 families affected with specific language impairment. We found that FOXP2 binds to and dramatically down-regulates CNTNAP2, a gene that encodes a neurexin and is expressed in the developing human cortex. On analyzing CNTNAP2 polymorphisms in children with typical specific language impairment, we detected significant quantitative associations with nonsense-word repetition, a heritable behavioral marker of this disorder.
      Vernes et al. 2008

Michelle quickly scanned the slide. "OK, all the main points are there, but I need to interpret the jargon. Basically what it says is that they found mutations called single nucleotide polymorphisms, or SNPs in the FOXP2 transcription factor. That factor controls a gene involved in brain development, and the mutations are correlated with speech impairment in children. That's why we think the human version of FOXP2 played such an important role in evolution. It is a language enabler. It permitted early humans to develop a primitive language, which was such a selective advantage that human beings have now become the dominant primate species on Earth."

All eyes fixed on Michelle. Several hands immediately shot up, but she looked at her watch and then over at Koskin. "Don, can you take it from here? I'm afraid I have a board meeting at TransTek now. I'll take one question, then I have to go."

Arthur Epanchin got her nod. "If the FOXP2 gene did this, why haven't other FOX genes done something equally or even more important? And don't we now have the ability to create FOX genes that would improve our species?"

Michelle looked amused. "That requires a very long answer, more time than I have just now. We're trying to do something much simpler over at TransTek, but you have raised probably the most important ethical question that faces geneticists today. Maybe we should do more than just fixing genetic diseases like cystic fibrosis or hemophilia. If we can manipulate genes, should we make our children more athletic? Better looking? Even smarter? We don't have time to discuss it today, but think about it. Years from now, you might need to decide."

Michelle turned to Don and said, "Sorry I can't stay, but your students can send me email if they have other questions."

Chapter 10

A smattering of applause accompanied Michelle as she picked up her briefcase and slipped out of the classroom. Koskin thought about something a golf pro had once told him, to quit practicing when you make the perfect shot.

Michelle had made the perfect shot, and he would not try to do better. "Let's wrap it up early today," Koskin said. "I'll see you on Friday, and be sure to check out those two papers Professor Murphy recommended. If there are any last questions, I'll stick around for a few minutes."

Most of the students closed their laptops and slipped them into bags and backpacks, then filed out of the classroom. Jim Battey and Julie Flanagan remained seated, waiting while Koskin unplugged the iPad and zipped it into his briefcase. Jim had a simple question about the midterm, and Koskin explained that it would be a mix of multiple choice, short answers, and one essay.

After Jim left, Julie stood and asked, "Professor, when are your office hours?"

"Right after class. Do you know where my office is?"

Julie shook her head.

"Walk along with me, because that's where I'm headed. Is it a short question?"

"Actually, it's kind of complicated."

They walked to the elevator down the hall, and Koskin pushed the up button.

"Julie, I remember Dr. Adamski recommending you, but I don't recall your last name."

"It's Flanagan."

"I used to know a Sharon Flanagan up in Portland. Any relation?"

"Mmmmm, not that I know of. I'm from Ashland, but there's lots of Flanagans in the Portland telephone directory. Flanagans are good at reproduction. One of my brothers lives there, and he has six kids so far."

The elevator arrived, the doors opened and a synthetic voice announced, "First floor. Going up," with a rising inflection on the 'up' as if 'up' were a place. They entered and Koskin pressed the button for the fifth floor. They rode up in silence, until the voice said, "Fifth floor. Going down."

Julie laughed as they turned left and walked down the hall. "Hearing that every day would drive me crazy."

"What? The elevator? I usually take the stairs for exercise. Sitting is the riskiest activity scientists engage in. Here we are."

Koskin keyed his office door open and the lights turned on automatically. "Have a seat over there on the couch. I'll just be a minute."

Koskin had received the double sized corner office as a perk when he became department chair several years earlier. Julie sat on the black leather couch and looked around. Several bright and bold abstract paintings hung above the couch, slashes of primary colors. One wall was a bookcase completely filled with books. The espresso machine sat on a small table in one corner. "Nice!" she said. "Who did the paintings?"

"That would be my wife. She's a graphic artist, mostly computational these days, but still enjoys using oil and acrylic paint on canvas when the spirit moves her. I'm making an espresso—want one?"

"Oh, I don't think so. I already had coffee this morning, and too much caffeine makes me jittery."

Koskin tamped the finely ground espresso coffee into a small metal cup which he fastened to the machine with a twist. After pushing the on button, he returned to his desk and sat down. "Okay, Julie, you said it's complicated. I'm listening."

"You probably know who Art Adamski is."

"I've read his papers, of course, but don't know him personally. Our only correspondence was the note from him recommending you for my course."

"Well, he certainly knows you, or at least your research. I went on digs with him for several summers, and when I came here, and he asked me to show you something."

"Wait a sec, Julie. My espresso is done." Koskin carried a small cup to the machine, filled it with the strong, black brew, and sat down again.

He looked intently at Julie, as though seeing her for the first time. "Adamski wrote that you were on site with him—Southern Ethiopia? That's impressive."

"That's right. It's called Awashi Valley. We did a preliminary survey last year in a dry river bed and found bone fragments in a sedimentary layer dated to 190,000 CE, so this year we had permission to spend another 4 weeks there."

"Who found the fragments?"

"I did." Julie reached into her pocket and pulled out a small box. "And this year I found something else."

Koskin held out his hand, but Julie did not offer it to him. "Before I let you see this, I must tell you that it's confidential. Dr. Adamski said you would understand."

"I don't understand yet, but I promise not to say anything outside of this room."

Julie opened the box and lifted out a small glass vial with a coded label. "Please be careful. It's sterile and sealed under nitrogen." She passed it to Koskin, who examined it with interest.

"Hmmmm... First bicuspid. Looks human. Modern."

"Bicuspid yes, but not modern. We found dozens of teeth, most of them attached to a few partial skulls. This one was pretty loose, so I used forceps to remove it and put it in the vial. The skulls were mixed in with obvious baboon bones."

Koskin looked up at her. Surprised, but then running through the possibilities. "Baboons? They wouldn't mix socially, but they might die in the same river or its marshes. Flooding or scavengers could have brought them together."

"Except for the way the bones were mixed. One of the smaller human skulls, probably a child's, had obvious bite marks matching a baboon jaw. And we have a human fibula in a baboon jaw. I have pictures for you." She pulled her laptop out, booted and put it on the desk in front of Koskin. She stood by the side of the desk as he studied the pictures.

"We think that a group of baboons was feeding on human remains when they were caught in a flash flood. Everything got buried in sediment and then preserved as fossils."

Koskin was moving back and forth through the photographs. When he looked up she was waiting for his reaction.

"So, you think, Adamski thinks, and I'm beginning to think that if the 190,000 CE date is correct, this could be a tooth from one of the first true humans. Though a few other hominids were wandering around Africa at the same time, but maybe not with teeth like these."

Julie nodded. "Dr. Adamski wants you to see if there is any DNA left in it. Probably not enough for a complete genome, of course, but if we're lucky, we might find fragments with the base sequence of FOXP2."

Koskin sat back in his chair, looking down at the tooth. He reached for his espresso and finished it in one gulp. It had gotten cold. "Are you in contact with Adamski?"

Julie nodded. "I have his private cell phone number."

"Call him right now, tell him the answer is yes. In fact, tell him the answer is Y-E-S!"

Julie's face lit up with a huge grin.

"But before you call, there's one last thing. You found this, right?"

Julie nodded, wondering what was coming next.

"I'm going to need help in the lab. DNA sequencing is easy these days, but getting it out of the tooth will take time. You're a grad student here, so this could be part of your doctoral research. Would you like to join my group? I have an open RA salary on my grant."

Julie leaped to her feet and clapped her hands, mimicking Koskin. "Y-E-S!" She knew she was not being Julie "Cooler" Flanagan, but she said again, "Yes!"

Koskin smiled, and decided he was going to like his new grad student.

Chapter 11

2020 June, Astana, Kazakhstan

Long before Michelle and Avalon had checked into the hotel provided for them by Nazarbayev University's Center for Life Sciences, she had updated her past knowledge of the country, its history before and under the Soviet Union, its emergence as a newly independent oil giant, and its leadership. She read translations of recent government press releases and soaked up the tone of national web sites. President Nazarbayev had died just over a year ago and was now idealized as "The Father of Kazakhstan." With new elections scheduled for the fall, acting president Bair Ducinbekov, a protégé of Nazarbayev's, was struggling to gain name recognition in a crowded field of contenders. Most observers doubted he could hold onto power.

Although President Nursultan Nazarbayev had once been a typical communist boss, after inheriting the country he had announced that Kazakhstan would enter the first ranks of civilized countries by the year 2030. He said he intended to guide the country as Singapore's famous Lee Kuan Yu had transformed his once desperately poor country into a social, democratic, and economic miracle. President Nazarbayev also became a billionaire whose regime had managed to ease out most of the non-Kazakhs from leadership in business and government. The transformation was a non-violent program of ethnic cleansing, accompanied by a quiet, methodical, and thorough ethnic nationalism that included a somewhat fictional recreation of the Kazakh past. The future component was his "Kazakhstan 2030" goal promising first world living standards by 2030. Instead of sending the most talented students abroad, they would be trained at Nazarbayev University in the gleaming new capital, Astana, which meant that foreign professors were now coming to Kazakhstan. Michelle had the passing thought, "Am I being recruited?" The thought even crossed her mind that perhaps her host was hoping to recruit her as a spy, not on the US but on the biotech industry, on her own employer, TransTek.

As Michelle and Avalon queued up for passport inspection in the Astana airport, a large man in black leather jacket, black shirt and pants stepped up beside them and held up a sign with large computer printed letters, "Professor Murphy and Miss Avalon." He bowed slightly and said, "My name Kairat. Good Shepherd for Dr. Akenov. Please follow." Michelle was puzzled. Was this the Kairat who had driven her and Hank from the airport 13 years ago? She didn't think so. This one was older, fatter, tougher looking. They followed him to a young man in military uniform who waved them into a small office where he inspected and stamped their passports and in English said, "Enjoy your stay."

Kairat picked up their heavier bags from the baggage area and with one in each hand nodded to the customs inspector and walked out of the airport with Michelle and Avalon following. "Wow. VIP treatment," Avalon whispered to Michelle, who nodded and smiled at her. Kairat put their bags into a large gleaming black Nissan Pathfinder sitting with the motor purring. A uniformed airport official standing guard nodded as Kairat held the door open for his two passengers, then got into the driver's seat and accelerated away from the curb.

On the drive from the airport to the hotel, Michelle was surprised at how much Astana had changed since her last brief visit on the way to Almaty. She remembered the remarkable Baiterek Tower in the city center, a 30 story white trellis with an enormous golden ball gleaming at the top. The story she heard was that it had been designed by President Nazarbayev himself to recall the Tree of Life legend about a sacred bird that laid its egg at the top of a tree every year. *Everyone has to have legends,* she thought to herself. She certainly had enjoyed the endless Irish legends her grandparents told as if they were history. She often regretted that science had to operate entirely without legends, though science was now old enough that some of its heroes and heroines were becoming myth and legend. The purpose of legend and myth, she thought, was to console and inspire. The purpose of science was to discover.

Kairat pointed out the Khan Shatyr shopping center which was new to her. It was shaped like a huge tent, perhaps reflecting Khazakhstan's original nomadic people. Then they passed an extraordinary 20 story tall glass pyramid, the Palace of Peace and Harmony, open to the people of Astana.

After nearly an hour of driving, made longer by congested traffic and their tour guide showing them the city sights, they finally reached the hotel. After Kairat transferred their bags to the young man in a black leather jacket and black pants who rushed out to help, Michelle offered him a few dollars, but he put his hand on his heart and said, "No thank you. Good Shepherd." Then he closed the door behind him and drove away.

Even after their long flight, Avalon was full of energy and eager to explore her new surroundings. As soon as they had washed up, Michelle spread on the table the contents of a thick welcome envelope that had been waiting at the hotel desk. She breezed through a booklet titled in several languages, "Young Visitors' Program" and handed it to Avalon, saying, "I think your conference will be better than mine."

Avalon began to skim the pages. "And I don't even have to give a speech," she said. "Can I go try out the pool?"

Michelle watched Avalon slip into her one piece suit. No one could mistake this tall, thin, brown girl for her natural daughter. She had a fleeting fear that Avalon might want to stay here where she had been born. And would she meet boys who would find her attractive? Or more to the point, would they attract her? *Well,* she told herself, *it's supposed to work that way or genes would never get passed on.* Civilization at least has added a lot of preliminary ritual to draw out the decision. *And maybe that's what the final phase of parenting is supposed to focus on,* she thought.

Avalon draped the hotel robe over her shoulders and said, "Don't worry, Mom, I'll be back."

Michelle had grown used to Avalon's intuition being close to mind reading. Avalon's explanation was that she could "read people". Don Koskin had dubbed her talent the Santa Sense, quoting from the jingle, "she knows when you've been naughty or nice, so be good for goodness sake!" As Avalon opened the door to go, Michelle said, "Remember, men here are different than back home."

"Same to you, Mom," Avalon said cheerily, then added, "Besides, everyone seems to be different from me." She closed the door. Michelle didn't know whether to laugh or cry. Avalon's sense of isolation had been growing much stronger as puberty came on. Although Michelle was sure she would never be a "terrible teen", she was deeply worried now that Avalon might have a terrible teenage life.

Michelle was not surprised when she found a full color brochure in the program packet, printed in the blue and gold colors of Kazakhstan and praising the conference's organizer, the geneticist Dr. Arman Akenov. *The world famous geneticist,* Michelle mused. His face stared at her from the first inside page. From beneath a traditional cone shaped Kazakh felt hat his round face and brown eyes looked straight out. The pronounced widow's peak pointed aggressively down to a marked vertical crease on the forehead and between the eyes. The face bore a strange smile expressed by the mouth only. The eyes could be a dead man's. Michelle wondered if she would understand the man behind this false smile. Avalon probably could. She had recently become fascinated with face reading. She had learned every muscle of the face and watched hours of Paul Ekman videos. Michelle had learned she could no longer hide any emotion from Avalon.

Akenov's official biography began, "Even as a child" Dr. Akenov lived up to the meaning of his name—Arman or Dream. Kazakhs have always been superior horsemen and successful sheep herders, but almost as soon as the young Arman Akenov could walk, his understanding and control of animals seemed miraculous. He became known in the mountain pastures above Almaty as "the boy who talks to animals."

The biography went on to relate how his understanding had made him famous for curing sick animals, training horses, and understanding the needs of his father's sheep and how his family had prospered by his talents. Michelle wryly translated into Hollywood and TV speak, *The Sheep Whisperer.* The bio listed the author of the text as a writer on the history faculty of Abylai Khan University.

The unprinted fact was that Akenov himself had rewritten it so thoroughly that it was autobiography—selective autobiography. The story was true, but its details embellished and cleverly political in style, selection, and emphasis. The brochure ended by declaring that Arman Akenov's work reached far into the past and would forever change the future. The past he had changed by showing that from the Kazakh gene pool came not only the first horse breeders, but before them the first of the three waves of Asians who had crossed the Bering Land Bridge and eventually created the great civilizations of the Hopewell, Aztecs, Mayans, and Incas. The Mongols who had conquered and held the greatest contiguous territory in history from the Chinese coast to the Balkans of Europe also shared Kazakh genes. For the future, Arman Akenov's discoveries and talents would "restore the ancient glories of the Kazakh people and help the nation achieve the President's Dream of world leadership in 2030."

Michelle at first chuckled at the bold play on the word dream—the meaning of Akenov's first name in Kazakh—but at the same moment she recognized the dangerous arrogance. History, from the Egyptians and Sargon of Sumeria was full of dreamers of impossible dreams who had destroyed their own nations and murdered millions. Akenov was certainly an ethnic nationalist, but that in itself didn't make him a killer. Suddenly remembering the question in Don's class about improving the human race, she realized that was something that might interest a man like Akenov. She propped the brochure upright against a table lamp and settled into a well padded chair to stare into the face that stared back at her with its strange smile. She made a note to save the brochure for Evan Lipkovich.

The Akenov smile summoned her recall of recent articles about traits that child prodigies share with the autistic. She had an intuitive sense of this phenomenon when the 3 year old Avalon's intensity led her to look for the genetic markers of autism, finding instead the strange protein she still thought of as avalin. She folded the brochure and put it in a large envelope that she labeled "L&G" for Lipkovich from the CIA and Grable from the State Department.

Chapter 12

In Arman Akenov's earliest memory he is lying snugly bound from shoulder to toes in a light blanket staring up at the vivid blue sky through the center hole of the family yurt when a smiling woman's face comes between him and the sky. She looks into his eyes and pronounces the naming ceremony, "Cening ating Arman. Cening ating Arman. Cening ating Arman." Other women gather around, and his mother bends down and puts her lips on his forehead. He stares at her. He does not smile. She laughs and looks around at the others. "Ah, such a serious baby, like his father."

A few years later his parents worried when he did not grow as fast as other children his age. He seemed sturdy and strong enough, but both boys and girls began to look down on him and call him Stubby or Shorty. The names made no impression on him. He appeared to be immune to insults. And also to friendships.

When he was six, he watched his mother and her friends name another infant, his sister. "Cening ating Anara. Cening ating Anara. Cening ating Anara." He turned away and started for the door, but his mother called after him, "Arman! Come see your sister."

Arman said nothing but turned and faced the women who were looking at him. The serious baby had become a very serious boy. In his first month of school he had mastered reading and writing and math as if they were simple card tricks. In fact, he mastered anything that had a system, from tending his father's sheep and horses to mechanical toys.

But he hated surprises. They scared him. The sudden appearance of a stranger terrified him until he was almost ten. If he saw a stranger coming toward the yurt, he would jump on a donkey and ride into the hills. If he was too late to run away, he would hide behind one of the rugs hung on the walls of the yurt. One evening when he was seven a wolf approached the flock he was assigned to watch. Arman froze. The sheep ran, but the wolf caught a young lamb and after tearing its throat open, looked Arman in the eye and dragged away the carcass. Arman still could not move. He thought he had turned into a tree. When his arms and legs and neck began to bend to his will, he understood the difference between wild and tame. To tame something meant you had to kill its wildness. Wolves could not be tamed, so men killed them.

He understood how tame things worked. Their logic fit into his logic like pieces of metal flying to a magnet. He understood people this way too, although people were not as easy to understand as sheep and horses. But he learned quickly that they acted one way when their mouths turned up and their eyes narrowed and their teeth showed and they laughed. They acted another way when their lips went tight and their bodies tensed, and yet another way when water flowed from their eyes and their mouths puckered. He saw that when one person acted this way, others often acted the same way and made the same expressions. He understood that he was a person, but other people seemed more like sheep than like him. And as with the sheep, he recognized a system in their ways, and he had mastered their system fairly well.

So when his mother called to him to see his new sister, he knew from the tone of her voice and the smile on her face what she wanted, and he decided to look. He made the corners of his mouth turn up like hers and walked toward her. The women opened a way for him. He looked down at Anara lying in his mother's arms in the loosened nest of her swaddling blanket like a little monkey peering out of a caterpillar's cast off cocoon. She was a week old, and she gazed up at him, meeting his eyes the way babies do. He held his forefinger down near her tiny hand and was surprised when she grasped it firmly. Arman felt no love or affection, but he wondered whether her grasp could bear her weight. He pulled upward. She began to rise but lost her grip on his finger. Not much of a monkey, he thought to himself. He turned his face with its fixed smile toward his mother, then swept it over the women gathered around. His mother laughed and hugged him to her breast. He hated to be hugged. Her arms around him were electric but without a shock—an uncomfortable tingling that stayed with him for a long time. He learned to tolerate it by closing his eyes and trying to go away in his mind. He knew what she would say just before she released him, and she said it, "Arman, sweet, you are such a dear, such a dear little boy."

Arman hated that word *little*. And he wasn't sweet. Fruit and sugar were sweet. He was not. He determined he would not be little. From then on, he would hang for an hour or more by his hands from a tree limb hoping this would stretch his body. He even hung by his knees, but he remained small. As the difference between him and others his age became more obvious over the years, the more he determined to *do* big things.

He felt his mother's lips kissing him through the heavy black hair that covered his head and was satisfied with the effect of his smile. After she released him Arman made his way through the group, smelling their female scent, keeping the smile expression on his face. He glanced back and was satisfied that they had forgotten him and were now passing Anara around, cooing as women did. He pushed open the flimsy wooden door and stepped into the bright sunshine of late spring. Below him and far away the capital city of Almaty sprawled in a gray haze of coal smoke from the power and heating plants and auto exhausts. From the pastures surrounding the yurt the Tien Shan mountains rose up in a gradual, dark green slope of firs, then in a steep, rugged, massive white wall of snow covered stone that stretched across the southern horizon separating Kazakhstan from Kyrgyzstan. A cool breeze was flowing down from the mountains. Arman drew in a deep breath, savoring the crisp air and its scent of snow. He put on his blue woolen cap against the chill.

In school Arman impressed his teachers as distant but brilliant, a boy who grasped everything instantly. He had an uncanny ability to know what would happen next in a story. He knew how to apply a mathematical formula in ways they had not thought of. He never broke the rules, but he broke a few teachers' hearts by refusing to respond to their jokes, or by embarrassing them with questions they could not answer, had not even thought of. He also irritated them by his arrogant dismissal of important historical figures, especially the

Kazakh batyrs, the heroes of his people's history and legend. For instance, he called Malai Batyr who had led the final campaign to defeat the Dzhungars of China a fool for not expecting that the gift saddle from his newly defeated enemies had been poisoned. The pointed dome of Malai Batyr's monument by the Kegen River, he said, was a dunce's cap.

He was equally scornful of the world's great generals and kings, dictators and tyrants alike. Yet he never went over the edge and made a claim he could not back up with facts and logic. He was meticulous about facts. No one knew more facts than Arman. He collected facts as if they were all rare stones and stamps. In his world the central fact of history was that Genghis Khan created an empire never equaled by any other power. Besides Genghis Khan, the only person he never criticized was the president of Kazakhstan. He had not yet read the section in the Kazakhstan constitution that forbade criticizing the president, but he knew from the way others spoke, that he should not cross this line.

His one strong enthusiasm in school, the one subject where teachers detected something like an actual passion, had been science. Especially biology. If he could be said to love anything, he loved the quiet time he spent in the pastures and the forests beyond, trying to understand the complex ways that trees, moss, lichens, fungi, birds, lizards, snails, mice, sunlight, rain, snow, grass and even the moon were part of a system. Several times he caught and killed a rabbit or a lizard with a stick. Then he methodically cut it open and studied the organs and their connections. He was always fascinated by the killing of a sheep. His father before cutting its throat would call him, knowing that he would want to inspect the organs and help separate the still warm heart, liver, and lungs.

One thing he could not understand was what his mother meant when she said she "loved" him or what his father meant when he said he was "proud" of him. He did understand, however, when his father told him, "you are a small boy for your age, but your mind is taller than most." That pleased him very much.

By his sixth school year he knew everything that would be taught for the rest of the usual 11 year education. He not only knew all that, but was applying what he learned—training dogs and horses, breeding them, designing his own experiments with everything from the hydraulics of the little stream that crossed the family's pastures to how sounds affected the behavior of sheep.

He also experimented on other children in ways that no one noticed. He quickly recognized in their behaviors many of the same behaviors he saw in sheep and horses and dogs and even birds. Children cared little about anything but hunger and thirst, boys pissing, tussling, learning to act tough; girls giggling at their antics. Arman did not play and he did not giggle. He watched other children play, and he saw they were like lambs and puppies.

One spring day when Arman was 10 years old he was watching boys from other families wrestling. He watched a girl sitting in the sunshine by her family's yurt combing another girl's hair. The boys were wearing no shirts and their bodies were glistening with sweat. They were young animals, and Arman easily imagined away the girls' clothes, and they too were young animals, one grooming the other. And here a thought occurred to him that changed his life forever. "I am not the same species," he said to himself. That was why he was small. He knew it was true. He was a serious boy.

He knew he was not as strong as those boys wrestling, but then again, he was not as strong as a cow or horse. But he was far more intelligent than any of them. In fact, he was more intelligent than anyone he knew, including all of his teachers and other adults he had met. He found a way to prove it. He chose the school bully, Bulat. This big boy often "bumped into" Arman in the school halls and said, "Watch where you're going stumpy." One day on the playground he had kicked a soccer ball like a rocket into Arman's face and made blood run from his nose. The bully Bulat often found Arman during lunch break, inspected the food he had brought, and he took what he wanted, sometimes all of it. He'd say, "A little guy like you doesn't need all this food." Arman's mother would wonder why Arman drank all the kumiz (mare's milk) she gave him at home but no longer took it to school. Bulat stole it. He also called Arman Sissy and Sheep Screwer. While Arman sat thinking with his back to a tree, Bulat came up behind him and pissed on him. Bulat was an animal. Bulat was wild.

Then one day Arman asked his mother for kumiz to take to school. She was pleased and dipped enough out of the big churn skin to fill his bottle. "I will be very careful, mother," he promised. He had told the truth. When he found a place to be alone on the way to school, he poured out some of the white and watery kumiz and replaced it with the fluids he had extracted from the *poganka* or bad mushrooms that grew among the first above the pasture. For good measure, he topped it up by pissing in the bottle. Kumiz had a little tang anyway. When lunch break came, he went outside as he often did to eat what his mother had sent with him. He had held up the bottle of kumiz at the right time to be sure Bulat saw it. When Bulat came to inspect his food, Arman said, "Is this what you want?" He held up the bottle of kumiz.

"Now you are getting smart, Shorty," Bulat said. He took the bottle and paused before he walked off. "Maybe the teachers are right," he said. "You really do learn fast. I'm getting to like you. We might even get to be friends."

Arman said politely, "I hope it's not too late." This time his smile included his eyes and his face felt very warm.

As they crowded back into the building for afternoon classes, Bulat caught up to Arman and stopped him with an arm on his shoulder. "Bring more tomorrow. That had a special flavor." Arman turned the smile up toward him. "I pissed in it," he said.

Bulat laughed and said, "If you did, I'd break you in half and make two stumps. Understand Stumpy?"

"Of course," Arman replied. "I understand."

He understood very well why Bulat was not at school the next morning, even before word came that Bulat had been rushed by emergency ambulance to the hospital in Almaty. And the next day when word came that Bulat had died from paralysis of the lungs and chest, unable to breathe, Arman also understood. When the hospital reported that doctors believed Bulat had died from mushroom poisoning, the teachers lectured all the students on the dangers of poganka. Arman had observed, "Bulat said no mushroom could kill him, and he would eat them to prove it. He was wrong."

The next day on the school bulletin board appeared a large drawing, an excellent likeness of Bulat's face looking very peaceful. Under the picture in fine lettering were the words,

БУЛАТ, КОНЕЦ, ДОСВИДАНИЯ [Bulat, the end, good-bye]

The artist had signed it in the lower right corner, *Чабан* [*The Shepherd*]. Students praised the drawing—so like Bulat—and they asked each other, *Who is this Shepherd?*

Arman decided he would begin keeping a journal of his experiments. He invented a secret language written vertically as in Chinese but read from bottom to top, beginning with the first column on the right. He wrote in wonderfully ornate script that looked like Arabic but was actually methodically embellished Cyrillic letters reversed. As a left hander he had long ago realized he could write backward as easily as forward.

His journal began with a title in large letters of his personal alphabet,
DIFFERENT
AM
I

On the next page he wrote the following verse with the title "POWER".
High in the mountains white--
Snow leopard.
Here in the pastures bright--
I'm shepherd.

From that time on he thought of himself as "The Shepherd." Now and then in school another student would receive a note with a cryptic rhyme or a stern warning, signed by "The Shepherd." Sometimes on a tree or on the wall of a corridor or on a bulletin board appeared a cartoon mocking a teacher or a student, signed by "The Shepherd." Everyone had an opinion who this Shepherd was, but no one guessed Arman. Arman could not make a joke or play a prank.

Arman was finished with school 3 years earlier than others his age. By his 9th year his teachers and the rector said he was ready to move on. In fact, they were also glad to see him go because whenever he was present in a classroom, he would ask questions no one could answer, or he would add information that teachers had not learned. In short, they were all glad that when he was gone they would be masters of their rooms again. The rector prevailed on his friend, the dean of Almaty's Institute of Agriculture, to offer Arman a place. Two years later Arman had finished the required courses, each with the highest grade, a 5, and he became their youngest graduate with the added honor of a Diploma of Distinction. He chose to stay on and work in a genetics laboratory headed by an old and slightly dotty Russian who was more than happy to have Arman restore the prestige of his lab.

Soviet biological sciences were only then beginning to recover from the long political ban on work that contradicted the belief validated by Lysenkoism, that a new society like Communism could create a genuinely "New Man" as Lenin had promised. During Perestroika, the re-building of Soviet institutions led by President Gorbachev, Arman learned enough English to read scientific journals, and he devoured everything he could find in western publications. Everyone assumed he would soon win the title "Hero of the Soviet Union." In August 1991, of course, the Soviet Union collapsed and its 15 "republics" declared independence.

In the first year of Kazakhstan's independence Arman was quick to claim a place in schemes and dreams for the future. Both the dreams and the schemes were grandiose, especially among the country's Kazakh population. For reasons both founded and unfounded many Kazakhs resented the near majority of Russians in the country. Many of the Russians, for reasons both founded and unfounded, believed their culture, education and intelligence was superior to Kazakhs. The Soviet regime had been clever enough to appoint the popular but obedient Kazakh, Kunaev, as leader of the republic and to maintain his popularity by endowing him with favors for his subjects. In 1986, however, Gorbachev's Perestroika (re-building) replaced the apparatchik Kunaev with a Russian. The smoldering resentment Kazakhs bore against Russian rulers ignited. They went into the streets to protest and their protests turned violent. The Soviet regime called out Russian workers and troops. Several Kazakhs became "martyrs." When peace returned, the Russian ethnics were wary and the Kazakhs resentful. At independence in 1991 the Kazakh president Nursultan Narzarbayev, a former steel mill worker, quickly put himself in charge of the new country. Arman knew that from this point on unprecedented opportunities would be available for Kazakhs like himself.

Arman used his contacts to propose a new Kazakh Institute for Biology of the Future that would be announced as a replacement for the old Soviet biological warfare testing facilities. Its funding would come from Operation Sapphire, the American effort to collect and buy all the enriched uranium in Kazakhstan and to return the nuclear weapons on its territory to Russia. Within 2 years, with funds channeled by the US embassy through the U.S. Agency for International Development (USAID), Arman had command of several buildings in a small research "village" outside of Almaty that had been devoted to nuclear physics. He was buying the very best western laboratory equipment. He recruited for staff only the very best Kazakh students, plus a few Korean ethnics.

The Institute's charter, written by Arman and approved at the highest levels of government declared that its mission was, "To devote the best scientific minds of the Republic of Kazakhstan equipped with the most modern technology to improving the environment, agriculture, and health of the nation's citizens." The U.S. ambassador was in the front row of ribbon cutters at the Institute's inauguration. Later, his wife, the elegant former ballerina Moira Hudson, said to him, "The director of that place has a very strange smile and evil eyes."

Her husband answered, "Maybe so, but if he's as smart as they say he is, the Kazakhs will have the best sheep in the world."

In its early years, Arman's institute had only limited funds to support his breeding program. He solved the problem in a unique way. In his research on the history of domesticated animals in Kazakhstan, Akenov discovered that decorative sculptures of horses had been found in several grave sites over 2000 years old. One in particular was a beautiful running horse sculpted in gold, probably a medallion to be worn by an ancient king or his queen. He used half of his savings to buy 300 g of gold, then hired a local artist to produce six medallions using the running horse theme, each of them different from the original in some way. He found a Swiss dealer who was not averse to taking a 10 % share of the sales price when one by one, the medallions were put up for auction in the European, Japanese, South American and US antiquities market. Only once was the scam discovered, but by then Arman had 100 million Tenge in his account, more than sufficient for his needs.

His Institute began to produce and clone superior sheep and horses and grains. That was the easy work. Graduate students and young scientists conducted this research effort, aided occasionally by specially selected foreign students whose reputations and work Arman had studied in the literature and on line before recruiting them. Luring them to his Institute in a country they considered exotic and beautiful proved little challenge, especially when he could also offer generous research grants, comfortable quarters and salaries twice what they might get elsewhere, all paid for by lucrative contracts he negotiated with new Kazakhstani businesses. Almost half of his funding came from the social development funds or "offsets" required of foreign companies that won contracts for construction, energy projects, and military supplies. He preferred to get his money from business, since that invited less political snooping and interference. And the businesses that paid what amounted to bribes had no way to look into the work of his Institute. Since most of the country's large businesses were run by Kazakhs, Akenov carefully selected a few to share his dream of a superior Kazakh people. They in turn selected from their workers and friends a cadre of tough and obedient sympathizers always willing to run errands for their bosses in their service to the great Dr. Akenov. Arman called them his Shepherds.

Kazakhs maintain strong family ties, a legacy of nomadic herding that limited viable units to families and clans. The family loyalties also meant endless pleadings that this or that relative be hired. Akenov's personnel department countered such requests by the simple expediency of declaring all Kazakhs one family in which each should have the position according to his or her merits. The result was that the Institute was a Kazakh style meritocracy that required all prospective employees to pass a skills and competency test. Akenov made sure the people who administered the tests were honest, and he had hired a woman with superb public relations skills to console the failures and their families.

When his foreign students wrote and published papers, Akenov's name was always among the authors, usually with good reason. Even when his contributions and guidance had not been primary, a little subtle bribery—extra grant or travel money—worked with Europeans and Americans almost as well as it did with his own country's bureaucrats. This subtlety included always wearing a spotless white lab coat whose design had just a hint of the cut of a Kazakh clan leader's traditional robe. His reputation grew, and soon it was admired not only in Kazakhstan but abroad. An American magazine writing about the new capital and culture of Kazakhstan called him "the white knight of Kazakh science."

What was obvious to both foreigners and Kazakhs was that Akenov had a thorough and practical knowledge of how the economic and political system of his country worked, and he had no trouble making it work for him. One way he did that was by carefully staying out of politics. He knew that if he stayed out of politics, the politicians would not intrude in his research. He also understood that after one or two more terms, the president would step down, tire of office, or perhaps die. Building a reputation as a national and international figure was a good way to get called into politics when the stage was empty of competitors.

As Michelle stared into Akenov's eyes from the luxurious chair in her two room suite, she had the feeling that he was staring back at her. Ridiculous, of course, but she knew from reading the scientific papers from his institute and from a few colleagues and graduate students who had met him that he was more than a brilliant scientist and a clever politician. She had read many of the papers from Akenov's institute. He had placed himself and his colleagues at the very front of genetic research aimed at improving animals and plants important to the world economy.

He even claimed in public that he had created smarter animals. What he did not publicize was that 20 years ago, while carrying out research on the development of the nervous system, he had taken blood from himself and purified the DNA from the white cells. Back then the technology for doing a complete human genome was still being developed, but it was possible to isolate and sequence smaller portions. At first, he was simply curious about what his DNA sequences looked like when compared to those of other humans. But something caught his eye. Genomes differ between individuals by just one in a thousand bases, but there was a sequence in his genome that had several unusual SNPs. Amazingly, it was buried in the well known sequence on the Y chromosome called the Ghengisid marker, inherited from Genghis Khan who pillaged and raped and ruled across Asia 900 years ago. Genghis and his sons had sired hundreds of children who passed the DNA sequences in his Y chromosome down through generations of Asian men. Other human beings had bored Akenov since his teenage years. He was lonely, but at the same time he was not interested in the bother of a dull wife and family just to emulate Genghis Khan. He wanted to broadcast his genes. He began an experiment that, years later, brought Michelle and Avalon back to Khazakhstan.

As she read about Akenov, something in the totality of his work and his institute caused a tingle of suspicion in Michelle's mind. His agenda was larger than agriculture, larger even than improving human health. There was this hint of Kazakh superiority, a kind of racism, but she could see other explanations for the wording that gave this impression. Maybe he was just a twenty-first century eugenicist like American scientists and politicians who had supported "racial hygiene" and eugenics in the early twentieth century. She made a note to elaborate these thoughts and put them in the envelope marked L&G.

Maybe she should go check on Avalon at the pool, she thought. She was not sure she could stay awake. *Heck of a spy I am*, she mused. Suddenly she felt herself sucked into the sleep of jet lag of 14 time zones in 17 hours. She almost collapsed on the giant bed and slept in black oblivion for almost an hour before Avalon returned and woke her by standing beside the bed and reciting a few lines:

"Over the edge of the purple down,
Where the single lamplight gleams,
Know ye the road to the Merciful Town
That is hard by the Sea of Dreams. . ."

Michelle shook her head and blinked. "Kipling," Avalon said. "You were in Dreamland and I was in a dreamy pool. Maybe I'll swim my way through the weekend." She paused while Michelle sat up and rubbed the sleep out of her eyes. "And, no, I didn't meet anyone dreamy. Mostly little kids with their parents and a few adults who came for the conference."

"Do you feel like a native here?" Michelle asked.

"I can tell I look like one, but besides some decorations, this hotel isn't Kazakhstan, unless Kazakhstan is some kind of United Nations."

"This is a new capital and from what I've read the architecture and planning are pretty international, so most of the new city is less than 20 years old. It's not Kazakhstan. Maybe it's what Kazakhstan wants to be."

"I'm learning some Kazakh," Avalon said cheerily. "I should, shouldn't I? A few people assumed I already spoke it because I look Kazakh."

"I wish I had your ease with languages," Michelle said. Avalon already spoke Spanish and did quite well both reading and speaking Mandarin Chinese. "But I think if you want another language, try Turkish since it's the root for the Central Asian languages."

"Rakhmet," said Avalon.

"You're welcome," said Michelle, "but I can't remember that in Kazakh."

"Okasi zhok," Avalon supplied.

Michelle sat up, then stood up and stretched and murmured a few lines from Frost:

"The woods are lovely dark and deep
But I have promises to keep
And now I've had my little sleep."

Chapter 13

2020, Eugene, Oregon

Over breakfast, Jodie asked Don whether he had heard from Michelle. Don shrugged. "Don't know. She's only been in Astana a couple of days, and might not have easy access to the internet. But I'll bet they have better weather than we do right now."

Koskin was not superstitious, but his optimism and pessimism tended to sync with the weather. The Willamette Valley didn't have a real winter where snow crunched under foot and freezing air scrubbed the face, and it didn't have torrid or sweltering summers—it had a sunny, warm, dry season and a wet chilly season. Today, however, summer had taken a rare absence. He had looked briefly at the weather maps. A fat band of clouds angling down out of the Bering Sea had leaned unusually far inland and south, bringing lightning and intermittent downpours.

After showering and gathering up his briefcase, Don kissed Jodie on the way out the door and set out for Starbucks during a pause in the rain, walking as fast as most people jog. For a block his mind wandered to bands of brown people in furs crossing the Bering Land Bridge over 15,000 years ago. Or had they skimmed along the seafood-rich coast in skin boats? Hundreds of them, perhaps even thousands must have trekked to North America as they slowly made their way south into a vast unknown land. They didn't know where they were going, but he did. This afternoon everything seemed possible.

Overall *Homo sapiens* had done pretty well, at least for themselves. Sure, other species, from dinosaurs to cockroaches, had survived thousands of times longer than *Homo sapiens* had existed, but none had spread so far so fast or changed so much in such magnitude. He sang to himself the line that the rock group Hair had borrowed from Shakespeare, "What a piece of work is man ..." He was about to debate himself on whether the line was sarcastic, serious or both when the rain came again, and he made a dash for Starbucks.

Julie and Don had agreed to meet there to discuss her progress in analyzing the DNA of baboons and the African tooth she had brought back from the dig. The remaining research effort to complete her PhD was to compare the ancient human DNA with the DNA of a modern human. No matter what she had to show him today, last week she had all but locked up the proof that her African tooth would make history and her reputation.

Koskin and Julie arrived simultaneously and found a table outside under the awning. As they waited for her cappuccino and his espresso, he leaned back and said, "Haven't we evolved nicely?"

The comment surprised her and she replied with a puzzled laugh.

"Never mind," Koskin smiled. "Just being silly. Let's get down to business. First, I want to touch base on the method you used to extract DNA from the teeth." With that, he pulled his iPad out of his briefcase, slid his chair around toward hers, and clicked on a folder titled Methods, then on DNA Extraction. A new screen popped up, and Julie leaned forward to read. The text was like a step by step recipe, except that the end product was not a cake, but instead a few micrograms of DNA. Julie skimmed down the list. "Yup, that's my Cordon Bleu culinary institute recipe. Even though I usually fry chicken and make sugar cookies, I followed it meticulously, and the results might surprise even you."

"Before we get into that, let's go step by step through the recipe."

"Hey! I think I just hit a home run and you want to review my batting posture?"

Don laughed. "Very briefly."

"Okay. Number one, I did everything under a laminar flow hood to avoid contamination by . . ." (her voice changed to a spook spoof) "the bugs that are everywhere unseen. *And* . . ." she emphasized the conjunction, "I sanded the teeth for the same reason. *And* I sterilized everything with bleach and ultraviolet. *And* you were right about that expensive dextran stuff. After I ground up the teeth to release any stuck DNA it was the dextran blue that helped me see the tiny bit of DNA that came out in it. Anything else to worry about?"

Julie paused, sipped her coffee and looked over at a table of laughing students who were getting up to leave.

Koskin studied her profile. She was no longer a student. She was a colleague, as good as any scientist he knew. She had done everything with perfection. Only a few months ago he had thought she might be too immature to work in his lab and perhaps had underestimated the difficulty of the research that might evolve into her Ph.D. thesis. He had given her a baboon's tooth to see whether she could handle the intense work effort. The tooth had been given to him by a colleague from Cardiff University in Wales who had worked in a catacomb under the Saqqara desert in Egypt where several thousand mummified dogs and a few baboons, cats, and other animals had recently been discovered.

A month earlier, Julie had passed her qualifying examination and was ready to complete her doctoral research. The exam is the last formal test for PhD students and is often called the oral because four faculty members engage in a two hour conversation with the student. Because Koskin was her research advisor, he could not be present, but Michelle Murphy had agreed to chair the committee, with three other members from the departments of Cell and Molecular Biology, Genetics, and Evolution.

Julie had prepared a 20 page research proposal based on her analysis of the baboon tooth. She had spent the first half hour describing her results and telling how she planned to study the teeth collected from mummified Egyptian baboons to see how much genetic variability there was in baboon populations 3500 years ago. She would compare their genomes with the published genomes from modern baboons and look for evolutionary changes that may have occurred in the intervening years. This was a perfectly valid research plan, but was a minor subterfuge because along with this work she would analyze the DNA in the 200,000 year old human tooth. She and Koskin had agreed with

Adamski's recommendation that the study be kept confidential until they had obtained the complete genome, so she had not included that work in her plan.

After her presentation, the faculty had grilled her for another 90 min, asking both basic and detailed questions. Some students wilt under this barrage, but Julie had passed with ease, feeling increasingly confident because it became apparent that the faculty members were learning from her, not the other way around. This is the best outcome of an oral. Instead of the usual tedious round of questions and answers, she could engage in a true conversation with her examiners, enjoying the essential creative activity of discussing possibilities and considering critical tests of the ideas being generated.

Then Professor Murphy had asked a question that startled Julie but led to the surprising result she was waiting impatiently to tell Koskin as they sat in Starbucks.

"Have you considered looking for point mutations in any of the FOX transcription factors?"

Julie had stared at Michelle for a moment, then answered, "Not specifically I suppose. But I will be looking for snips throughout the genome, and I imagine some might show up in the FOX genes."

Michelle had then asked a question in a way that was often used in oral exams. "If someday you are teaching an undergraduate class, how would you define a snip?"

Julie smiled, happy that she could answer such a basic question. "Well, that's just a word we use as shorthand for single nucleotide polymorphism, abbreviated SNP. At least in the human genome, there is approximately one snip per 1000 nucleotides, which means that humans are 99.9 % the same, but not identical, unless of course you are comparing twins."

Murphy nodded. "Now tell us what a transcription factor is, and why FOX genes are important."

Julie had paused to think. This was a little more challenging, but then she recalled the class in which Michelle had described the FOXP2 gene. She had taken careful notes and reviewed them just a week ago as she prepared for the oral, so she could lapse into fairly technical language. "Transcription factors are small proteins that control the transcription of DNA into messenger RNA," she began. "FOX is an abbreviation of 'forkhead box' and is used to denote a set of transcription factors that are particularly important in regulating embryonic development. Some of them are involved in controlling development of the nervous system in mammals, and one of them, FOXP2, is apparently essential for the neural functions that allow humans to develop and use language. It differs from the chimpanzee gene by just two SNPs, and is a wonderful example of how seemingly simple mutations can affect the course of evolution."

Murphy had smiled. "So you remember my talk in Professor Koskin's class?"

"I sure do," Julie replied.

Julie had celebrated the exam committee's unanimous decision with a 3 day weekend hike along the cliffs and beaches, letting the ocean winds blow through her hair and her mind. Then she returned to her lab bench, looking at a computer screen with Koskin standing next to her. They were going over her comparison of the ancient baboon genome with that of modern baboons. Five other students were in the lab, each with a bench space and a stool. Their desks were in a smaller adjacent room separated from the lab by glass windows. There they had computers that gave them access to a major computational facility used for bioinformatic analysis of base sequences in genomes.

The workhorse of the lab was a new nanopore instrument. To practice their technique, Koskin invited each of his students to scrape some cells off the inside of their cheek, extract the DNA by standard methods, then place 10 µl into the well of the fifth generation PromethION instrument. An hour later their genome was available for analysis, and they could download the sequence of three billion base pairs into a thumb drive. He reminded them that the first human genomes had taken 10 years and three billion dollars to determine, which just showed how much genomic technology had advanced since 2000. His students had fun comparing their gene sequences to published genomes such as those of Craig Venter and Jim Watson. They could also search for marker sequences. Two of the male graduate students discovered that they had inherited baldness genes from their maternal grandfathers (that was already becoming obvious) and one of the women found markers for a predisposition to diabetes, indicating that she should be careful with her future intake of carbohydrates.

Julie had proceeded from these routines to her baboon teeth. As expected, she had not found any major differences between the mummy teeth and the teeth of the modern yellow baboon. A few hundred generations had not been enough time for significant numbers of mutations to accumulate. She had been disappointed, but she had honed her skills very intently. Koskin suggested that it might be time to retrieve the humanoid tooth from the freezer and look for its DNA, and if she found any, to compare it to modern humans. The day she took the ancient tooth from the −80° freezer, Koskin surprised her with a glass vial containing a human child's tooth. "Present from Michelle Murphy," he had explained. "You can analyze her daughter Avalon's beautiful little baby tooth as your modern reference DNA. It's a nice internal control for you to use along with the public database."

Now that Julie knew what she was doing, the lab work had gone quickly. Most of the time required for bioinformatics research is spent doing the computational analysis in which newly established base sequences are compared to known sequences to look for differences. Julie was not sequencing a novel genome, so her effort was called re-sequencing. She had three teeth to work with, the one she had brought back from the dig, a second baboon tooth from Egypt, and Avalon's baby tooth. Over the next few days she followed the recipe established by other researchers for extracting DNA from ancient bones and teeth. The final result in the bottom of the centrifuge tube was a tiny blue dot barely visible to the eye that she stored in the lab freezer.

Julie's African tooth sample had suddenly become very valuable. A 200,000 year old human genome would easily have sufficient news value to be published in the top journals *Science* or *Nature*, and would assure that she would be invited to join the faculty at a major research university such as Berkeley, Harvard or UC Santa Cruz. Julie buried her sample at the bottom of the freezer so that none of the other grad students might accidentally happen upon it. She had been working sixteen hour days almost without noticing time, but with the DNA in the freezer and discovery imminent, she decided to relax all day Sunday. "Even God rested on the seventh day," she told a fellow student who expressed surprise.

Julie was so excited, however, that she couldn't stay away from the lab, which was dark and deserted when she arrived in the early evening. As she reached for the switch by the door, she said aloud, "Let there be light." It was her way of saying without jinxing her hope, *let the discovery begin.* At her lab bench, she added 50 microliters of salt solution to each tube and tapped it with a fingernail to disperse the blue pellet. She removed 1 microliter, added 1 μl of an enzyme called a helicase that would unwind the DNA double helix, then pipetted the almost invisible volume into ten disposable cartridges that were the core of the sequencing device. Each cartridge had a graphene membrane containing 10,000 nanoscopic pores, and each pore had an electrical connection and amplification that could detect currents measured in picoamperes. The device applied a voltage of 200 mV across the membrane which served to pull single molecules of DNA through the pores. As the enzyme fed the DNA into the pore at a thousand bases per second, the different shapes and properties of the bases produced distinct electrical signals as they passed through.

When Koskin had introduced Julie to the nanopore instrument, he had asked her to calculate how long it would take to run a typical human genome of three billion base pairs. She impressed him by doing the math in her head: a thousand bases per second times 100,000 pores equals 100 million bases per second times 3600 seconds per hour equals 360 billion bases per hour, more than enough for one genome. Even though it needed to be done at least 10 times to reduce the error rate, a 1 hour run would still provide plenty of base sequences for computational analysis of a very accurate genome. Julie had decided to do the baboon tooth first, her least valuable sample, then Avalon's tooth, and finally her African tooth on Thursday morning. She plugged the cartridge into the Promethion Sequencer, set it for an overnight run, then left the lab to finally get some sleep.

Monday morning, when Koskin had arrived at his office he found a yellow Post It note stuck on his door. "Come see! I'm at Starbucks." He hurried through a few short cuts between buildings and at the edge-of-campus Starbucks he found Julie sitting with her laptop, looking at rows of A's, G's C's and T's on the screen. It was still too early for other students, so they had a corner table to themselves where they could converse without being overheard. Don was already so pleased with Julie's work that seeing her overflowing with confidence this rainy morning he insisted on treating her to the coffee of her dreams with any pastry she wanted.

"I could drink African puddle water, and I wouldn't know the difference this morning," Julie said. "Order anything."

When Don returned with scones and a bear paw and fruit and two double lattes, Julie slid her chair beside his so he could see her screen.

"So, Julie," Koskin said, "What am I looking at?"

Julie pointed to the screen. "I did the other baboon tooth, Avalon's baby tooth and my African tooth. It will take several months for a complete analysis, but a few things are showing up. Most important is that there is intact DNA in my African tooth!" Julie turned and grinned at Koskin, who gave her the thumbs up and said, "High five!" They slapped palms.

Koskin looked at the screen where short reads were superimposed on a much longer string of letters of the reference genome. "Tell me about it."

"Well, the DNA is fragmented, of course, maybe 50 bases long on average, but there are alignments throughout the standard genome and that's what I was looking at. For instance, here's one in the cytochrome oxidase gene of the mitochondrial DNA."

Julie used a cursor to indicate a string of bases on top of the standard sequence which had one base different, marked in red. "Here's a snip."

"How many are you seeing?" Koskin asked.

"Oh, they're all over the place. I found quite a few in the Y chromosome region, by the way. That alone is going to tell an interesting story."

"So, the owner of the tooth was a male?"

"Oh, I should have told you that first. Yes, for sure."

Koskin paused, looking at the screen. "What about FOXP2?"

Julie glanced over at him with a huge grin. "I thought you'd ask that. Take a look."

She clicked on an icon on the desktop, and a new screen with several short sequences appeared, again superimposed on the standard human sequence. There were no SNPs. "It's a perfect match, at least the fragments I got out so far."

They both leaned back in their chairs, looking at each other. After a moment, Koskin said, "Wow."

Julie nodded. "The oldest Homo sapiens genome. We got it."

The two were silent for a moment, pondering the magnitude of Julie's discovery and all the new information that would emerge as she delved deeper into the comparative analysis. Then Koskin asked how far she had gotten with Avalon's genome.

Julie replied, "It's all done, but I haven't looked at the reads yet."

"I'm curious," Koskin said. "Can you do a quick scan for snips today? After that, run a BLAST search for FOXP2. It should show up on chromosome 7, so we don't need to do the whole genome."

"Why that gene?"

"Just a hunch. Avalon has tremendous language skills you know. And she can do other things most of us can't do. Maybe it has something to do with the position of FOXP2 on the chromosome, or perhaps there has been a gene duplication. Who knows? I'm just fishing. All you need to do is to run the software that will look for the FOXP2 markers in Avalon's tooth and compare the sequences with those in the public database. That will be fast."

Julie closed her laptop, took a last sip of her cafe latte and stood up. "See you after lunch."

Don spent the rest of the morning working on his review of Denisovan DNA in Asian genomes, then bought a muffin and coffee for lunch. He returned to his office, but when he had heard nothing from Julie by two in the afternoon, he decided to look in on her and see what was taking so long. He found her sitting at her desk in the grad student office space, staring intently at the screen of her computer.

"What's happening?" Koskin asked. "Any problems?"

Julie shook her head. "Take a look at this ORF." She was referring to an open reading frame indicating a sequence that would normally be translated into a protein.

Koskin sat down and immediately saw what had caught her attention. Usually when one human genome is compared to another almost no differences appear, so that rows of the four letters representing adenine, guanine, cytosine and thymine go on monotonously, with only an occasional snip showing up. If something doesn't fit, however, the computer displays it above the standard genome as a separate string of letters. Avalon's genome had a sequence that didn't fit.

"My god!" Koskin whispered. He took the mouse and immediately began to move the sequence across the screen, looking for familiar combinations of letters that he might recognize. He found one, a long string of CAGCAGCAGCAG... repeats of the basic CAG triplet, the code for glutamine. "It looks like a FOX gene!"

"Yes. I tried to match it to all the known FOX sequences, from FOXP1 to P4, but it's a new one. There are several scattered around on a sequence in chromosome 7, as though duplications have occurred."

Koskin leaned back in his chair, crossing his arms, staring into space as the implications sank in. Then he turned to Julie, and they locked eyes. "FOXP5. Let's keep this between you and me until Michelle and Avalon return from Kazakhstan."

Julie shrugged and said, "Sure."

Don looked at her, then said, "Listen carefully, please." He explained that Michelle had once looked for an autism marker and why she had refused Hank's insistence that they have a much more detailed genetic analysis. "She gave me this tooth and she now wants the sequence, but for sure we should be the ones to tell her about it."

He saw that Julie was nodding her understanding. Then he asked, "By the way, what was the sequence you mentioned that surrounded FOXP5."

Julie stared at Koskin for a moment, then said, "There must be some mistake, because it's unbelievable."

"Why?"

"A BLAST search calls it a Ghengisid marker."

Chapter 14

June, 2020, Astana

While Avalon dressed, Michelle stood at her windows and let her eyes wander over the Ishim River and beyond to the vast rolling steppe that was in climate and geology more a part of Siberia than Kazakhstan. In winter this land alternately lay under a blinding white field of snow or disappeared into blizzards that frequently buried cars and their passengers who had more hope than sense. The "forest" planted around the city over 20 years ago was still no more than 15 ft tall. Michelle's view in June, however, was a green and warm landscape and a city that was at once lavish, elegant, brash, and boastful.

Avalon came in from her room and asked, "How's this for the reception?" Her pearl silk blouse and brown slacks made two harmonized notes of color with her skin.

Michelle said, "Matisse could not have done better."

She had chosen her own clothes long ago—the blouse, skirt and jacket she had nominated for the reception, another ensemble for the keynote talk and the day that followed, a pants suit for the promised local tours, and two light and modestly slinky dresses for dinners and evenings. Her dress code was "blend in from afar; stand out on close inspection." The embroidery of wildflowers on the lapels of her dark green jacket was the same color as the jacket, but wonderfully intricate when seen close up. On her rich cream colored blouse she wore a tiny honey bee pin. Perhaps it went with the wildflowers. Maybe it warned she could sting. She enjoyed ambiguity. Sometimes she thought of herself as a Rorschach ink blot test. Sometimes she thought she was simply indecisive or didn't know what her message was. Both Michelle and Avalon pinned on the elegant blue tags with their names in gold letters in English, Kazakh, Russian and Chinese. In the lower right corner Michelle's badge identified her as "Speaker" and Avalon's as "Youth Program."

Avalon examined her badge and quoted Emily,

"How dreary—to be—Somebody!

How public—like a frog—

to tell one's name the livelong June,

to an admiring bog."

"Well, Miss Dickinson," Michelle replied, "I'm afraid you are going to be a somebody for the next two days. But instead of a croaking frog, you'll be a badged youth."

Over the doors to the reception hall a large banner boasted in several languages, "Welcome to Kazakhstan and the Future of Science." From a giant poster on the wall opposite the entry doors the strong, broad face of former President Nazarbayev beamed down on everyone with his kindest smile. A young Kazakh woman in a short skirt, high boots, and a traditional Kazakh cap with a vertical tuft of feathers read Michelle's name tag and bowed to her. "One moment please Professor Murphy," she said in nicely pronounced English. She held Michelle gently by the elbow as she called to another young woman who slipped quickly into the crowd. Turning to Michelle, she said, "Dr. Akenov wants to be the first to greet you."

For a moment Michelle thought the messenger was returning carrying a Kazakh conical hat before her, but under the hat she saw the face that she had studied on the conference brochure. Akenov was even shorter than Michelle, yet he walked confidently, parting the crowd as if he were six feet tall with the messenger woman trailing him respectfully. He wore a long tan camel hair robe that stopped just above his ankles. Although he was short, his proportions were normal. He stopped in front of Michelle and Avalon. He reached out for Michelle's offered hand, drew it to his lips and kissed her knuckles. "Call me Arman," he said, putting his right hand over his heart and bowing his head slightly in the traditional gestures of respect.

"Call me Michelle" she replied, "And this is my daughter, Avalon." Immediately she noticed that the man's mouth held a fixed smile but his eyes did not. In fact, she thought it odd that he only glanced at her face, and as he talked, his eyes looked at other things. Michelle saw that Avalon was studying Akenov's face, reading it, she was sure. She also noticed a man standing a few feet behind Akenov in the same black leather coat and black shirt and pants as the driver Kairat had worn. His face seemed pinched, his eyes wary as he watched them.

"Michelle and Avalon," Akenov said as if fixing their names to a bulletin board in his head with a slight jerky nod. "I look forward to getting to know you. You and Miss Avalon will be guests of honor at my table for tonight's banquet."

Again, Michelle noted that he did not look at either of them for longer than a fraction of a second. Michelle wondered, *Can he be autistic? Maybe Asbergers.* She knew the signs from her reading when she had worried that Avalon might have autistic traits.

Akenov told the waiting messenger to bring hors d'oeuvres and drinks. "Everything is from Kazakhstan. You must try our fine wine or champagne, and for Miss Avalon perhaps our finest kumiz. I know you were born in Kazakhstan, so perhaps you tasted kumiz as baby."

"Do you think so?" Avalon asked. "Would a baby drink horse milk? It's fermented."

Arman nodded his head slowly. "You know this. Yes, it is fermented but very slightly, and mother would give baby only small amount. To help immune system. See, you look very healthy."

Michelle noticed that while his English was very good, he never used the articles "a" and "the". She knew they did not exist in Russian, and perhaps not in Kazakh, but he had gone to Russian schools and had worked with Russian colleagues. She took a flute of champagne from the server's tray, and Arman handed Avalon a miniature bowl with kumiz. Michelle told Arman that the champagne was indeed the best she had had anywhere. Avalon sipped cautiously the tangy white kumis. She thought it had a taste something like a soda made with weak milk.

"Yes, yes," Arman said. "by 2030, our magic year, you know, we shall have everything here that you know in America or Europe, and some things are already better."

Avalon had read about Akenov's breeding program and decided to display her knowledge. "And next year, one of your Kazakh horses could be a contender for the American triple crown of racing."

"Ah, you know about this. You might say it is my design. But she will not be contender, Miss Avalon. She will be winner. It's our secret. You are first to know. But please excuse me now, I must continue to welcome guests."

Michelle spent the next hour renewing old acquaintances and meeting friends of friends. Avalon met a young Kazakh boy named Elzhan and tried her Kazakh words on him. She gave her mother a "don't worry" look and the two drifted off, the boy eager to find out more about this American girl who looked Kazakh but spoke only a few words of the language.

When Michelle and Avalon were seated at Akenov's table, Avalon asked him playfully, "Can you guarantee we'll win if we bet on your horse?"

"No guarantee," he replied. "We made horse but ordinary woman made jockey."

Michelle asked, "What do you mean, you made the horse?"

"Ah. You have not read our papers on GMO equines?"

Michelle, now paying close attention, shook her head. Arman smiled. "I understand you are on TransTek scientific advisory board, so I think you might be interested. We have isolated three genes that control neuromuscular control and muscle strength in Kazakh horses. We inserted these into several ova with viral vector and were fortunate to produce female foal that is now exhibiting superior speed as 2 year old."

Michelle made a mental note to send an email to the CEO at TransTek. She had never heard of this part of Akenov's work with horses. She asked her first leading question. "So, have you bred her yet? Are the genes expressed in her offspring?"

Akenov laughed. "Ordinary breeding is too slow. We are taking short cut. She is being cloned, and we have nine mares carrying her identical offspring."

This revelation stunned Michelle, but she managed to hide her surprise.

Akenov added, "Koreans cloned dog. I cloned horse. For me it is easy. Maybe human too."

Michelle said, "Maybe human—you don't mean you have tried to clone a human?"

Akenov stared across the room, at nothing Michelle saw that stood out. His answer was, "To me it is not logical that everyone alarmed about cloning human, even more alarmed than about killing them. Clone one is big disaster, great scandal. Kill ten or one hundred, just sad news."

Michelle thought of several ways to explain this, but she focused on science not public opinion. "But would you clone a human?"

"Someone will," Akenov said. "Maybe someone has, you know." Akenov leaned close to her and said in a low voice, "I can tell you very interesting secret, but you must not tell."

Michelle could feel his breath in her ear and didn't like this intimacy, but her mission was to learn what was not known. She didn't turn to him but said under her breath, "I won't tell."

"You know KL-VS gene?"

Michelle nodded her head. "I've read about it." KL-VS had been described by two California scientists who had been looking for a gene that might be used to diagnose Alzheimers. Instead they found that people with the KL-VS had higher intelligence on average, and that the gene inserted in mice improved their learning ability.

Akenov said, "5 years ago, I found this. Two years ago I transferred this gene to myself. Secret, understand."

Michelle was astonished that he would claim something she knew was impossible. She had to show little or no surprise, because maybe he was testing her in some way. "But you were already very smart, and KL-VS does not seem to have a big effect."

"It works," Akenov whispered. "Secret. Tell no one."

Michelle was skeptical, but nodded, wondering why her host would reveal something like this to her. If true, it would be a discovery worthy of the Nobel Prize. Waitresses, young Kazakh women, came by to fill vodka glasses for the first toast. The ceremonies began. Michelle stared at the waitresses, all tall and thin like Avalon, but a bit odd in their short skirts with knee high boots, at least to her Western fashion sense. *Half horsemen, half hostess, she mused.* She looked at Avalon, who was now seated across the table with Elzhan. Avalon would grow into that same mold. Arman must have been thinking the same thing. He caught Avalon's attention, then nodded toward one of the servers and said, "Someday, you will be as beautiful. Once Kazakh, forever Kazakh."

Avalon looked puzzled, embarrassed, and Michelle broke in. "You are a passionate patriot."

Akenov replied, "Yes. And I hope I am much more."

Chapter 15

Next morning, the keynote session began with the Kazakh national anthem, sung by a woman in traditional dress. She had a powerful operatic voice and a corresponding emotion on her face. A dozen musicians on stage accompanied her, playing the long narrow bodied, two stringed dumbros, the bass voices of the korbuz strung with horse hair, and several varieties of pipes and drums. As the anthem played, its words in several languages appeared on a large screen above the musicians. In the anthem's wake Akenov rose and walked to the center of the stage with a spotlight playing on him. He wore his ankle length robe of brown camel hair with light embroidery on the lapels and cuffs. He put his hand over his heart and bowed to the audience.

After the applause died, the screen above him again began to translate as he acknowledged with lavish and predictable praise, first, former President Nazarbayev who was present only in the grand picture, then acting President Ducinbekov, and then, descending the ladder of importance, government ministers, other scientists and on to his own staff and finally the staff of the hotel. He concluded by noting that just 30 years ago Kazakhstan had the humble position as the Soviet Union's nuclear testing ground and its rocket launching pad, and the Kazakh people were often mocked as nomads and sheepherders. He repeated the word for herder several times with different inflections—*chaban, chaban, chaban.* "Yes, it is true that the chaban often could not read or write. My parents could not read or write. But you know what the chaban does well—he protects and he leads. He knows where the wolves are and when they come. He knows where and when the pastures are greenest and most nourishing. I was one of those sheepherders. I am glad to be *chaban.* I am proud to be President of our Good Shepherds Foundation, and I am proud to be scientist. All scientists are *chaban*—scientists know where to find the knowledge that feeds civilization. With my country, with my people, and now with you, I am here at this conference where we gather as scientists, as equals from east and west, north and south."

As he spoke he opened his robe to reveal his white laboratory coat. "I am a Kazakh first, and then scientist." A murmur of laughter rose from the audience. He took off the robe and gave it to a young Kazakh woman who appeared suddenly at his side and carried it away. He continued, "I welcome you to my ancient and modern country—Kazakhstan. Once upon a time most of you had never heard of Kazakhstan. Today you are here. This conference is all about future that we will all share together. So it is fitting that our keynote speaker is not only outstanding scientist from half way around world, but she has come here with her talented and beautiful daughter who 13 brief years ago was born here in Kazakhstan and at this very moment is participating in our youth program. Now please welcome an outstanding scientist and mother, Professor Michelle Murphy."

Michelle walked confidently to the podium as the audience clapped politely. When she had entered the lecture hall a few minutes earlier, she had been surprised to see the number of people at the conference, over 500 participants. She was used to being one of the few women at international meetings, but here nearly half the speakers were women, although she was one of just a few westerners. She had often presented research talks to even larger groups at previous international meetings, and was comfortable on stage. Furthermore, at this meeting she was the keynote speaker, responsible for getting the conference off to a good start. The theme of the conference was genetics, evolution and the brain with several breakout sessions that included techniques for brain research, the importance of rapidly growing genetic databases, new technologies, and how certain diseases, particularly Alzheimer's affect intelligence by damaging the brain's nervous system. Michelle had decided to start at the beginning—ideas about the evolution of consciousness, and how she would differentiate between consciousness, intelligence, awareness and self-awareness.

When the applause had died, Michelle thanked Akenov and the audience, and clicked the control for the first slide which appeared brightly on the huge screen over her head. With her slides on the screen, the non-English speaking audience donned headsets that plugged into the arms of their seats.

"*When you measure what you are speaking about, and express it in numbers, you know something about it; but when you cannot measure it, your knowledge is of a meagre and unsatisfactory kind.*" Lord Kelvin, 1888

Michelle allowed her audience to read the slide or listen to the translation through their earphones. "Sir William Thomson, given the title Lord Kelvin after the river that flows through his home town of Glasgow, was a physicist and engineer in the nineteenth century. He received the highest scientific honor, even greater than the Nobel Prize, when his name became the unit of absolute temperature expressed as degrees Kelvin. Numbers served him well when he measured temperature with his thermometers, but what if he had been asked to measure consciousness? He would have replied that since we can't measure it, consciousness cannot be studied scientifically."

She paused for a moment to let this challenging statement sink in. This conference, after all, was basically about the science of consciousness. "On the other hand, you might know that Kelvin was notoriously wrong when he calculated the age of the earth to be between 20 and 40 million years, based on the rate at which heat is lost from a molten object. This was much younger than Darwin's approximation of 300 million years, which he estimated from the rates of erosion of a deep valley in England. Both were far off, of course, because radioactivity had not been discovered yet. From the rate of radioactive decay at which uranium turns to lead, we now know that the Earth is 4.6 billion years old—give or take a few million.

"Darwin may have been mistaken about the age of the Earth, but his theory of evolution by natural selection has stood the test of time. Today I want to use Darwin's brilliant insight as a way to measure consciousness. In my view, an experimental approach to understanding consciousness will treat it as the product of an evolutionary process leading to a threshold complexity of the nervous system that we can express as a quantity. We know that the first brain-like collections of nerve cells must have been very simple, yet half a billion years later the architecture of the conscious human brain is incredibly complex. If we consider consciousness a product of evolution, then over time complexity grew in increments until at a very high level of complexity consciousness appeared. My purpose today is to describe a quantitative approach to define the gradation and fulfill Lord Kelvin's challenge."

Michelle proceeded to describe the evolution of the first single and multicellular organisms, then the appearance of neurons and nervous systems. As her slides flashed one after another on the screen, she sensed the audience paying close attention. "During the Cambrian radiation between 580 and 500 million years ago, the first complex animals appeared that are now called *Bilateria*. It is reasonable to assume that these organisms had nervous systems. No doubt it would have been a major selective advantage for the predators and prey of that era to be able to sense their environment and respond with appropriate behavior. The chief characteristic of this level of nervous function is that the response to variable sensory inputs would have been a *reflexive sensory-motor response* with minimal modulation. This basic function is preserved in higher organisms in the spinal reflex we have all experienced in response to a painful stimulus. If you accidentally touch a hot surface, no one pauses to think, "Oh oh, that's hot. I should move my hand."

She waited for the translation, and was awarded by a ripple of laughter from the audience. "Instead," she continued, "the heat sensing nerve cells in your fingers send a pain message to the spinal cord, which immediately sends a motor signal to the muscles in your arm, causing them to contract and remove your hand from the hot stove. Keep that in mind next time you try to swat a fly. The fly never thinks about the descending fly swatter. Eyes see something coming, wings get the fly going—no thinking in between. It goes through its life using the equivalent of spinal reflexes."

"The next step after reflex is awareness. When we observe the behavior of fishes, reptiles, birds and mammals, it is clear that the vertebrate animals can modulate their nervous system in response to changes in the environment, and I define this response with the word *awareness*. Instead of being entirely reflexive, awareness allows animals to modulate their behavior within certain limits. The modulated response seems to reflect a short term memory measured in seconds. However, at this level of nervous complexity intelligent behavior is still not possible. With rare exceptions, most birds and mammals are unable to match even the most minimal human intelligence in terms of problem-solving. They are aware from moment to moment, but they are not self-aware. Thus a bird fights with its own reflection in a window, and a dog barks at its image in a mirror, not recognizing images of themselves.

"The behavior that characterizes *self-awareness* arose in the increasingly complex nervous systems of primates, and also in other large-brained animals such as elephants and dolphins. A self-aware organism recognizes itself in a mirror, and *Homo neanderthalensis* 400,000 years ago would likely have had no difficulty passing this test. Self-awareness evolved into modern consciousness 200,000 years ago with the appearance of *Homo sapiens* in Africa. If a child from that era could somehow be transported forward in time to today's world, it would presumably be indistinguishable from other children in its ability to develop language and adapt to contemporary culture.

"The most striking property of a conscious human being is not just self-awareness, but to varying degrees human brains can indefinitely maintain an internal model of sensory input and manipulate the model in order to predict future outcomes. Short term memory is therefore not measured in seconds, but instead can be maintained throughout a problem-solving interval. I define the word *intelligence* as the relative ability of the conscious nervous system to perform such tasks. I would guess that most of the people in this room have had their intelligence quotient, or IQ measured, and I would be willing to bet that everyone here is well above average."

Michelle's compliment elicited an appreciative chuckle from the audience.

"I will now present a set of postulates that can be used to clarify the discussion of human consciousness. The postulates, taken together, also suggest experimental and observational tests of hypotheses related to consciousness. The first postulate is that consciousness will ultimately be understood in terms of ordinary chemical and physical laws. This postulate links consciousness directly to nervous processes in the brain. It arises from the principle of parsimony (Occam's Razor)—that the most likely explanations will rely on the least number of assumptions."

Michelle stepped out from behind the podium to mark the end of her first argument. "I conclude that because consciousness emerges from the interplay of physical and chemical laws, there must be a way to measure it. In the next part of my talk, I will show you how it can be done mathematically, and how the results predict the intelligence ranking of mammals, from mice to humans." She listened carefully for any ripples, maybe groans of dissent. Nothing. She returned to the lectern.

Over the next half hour, Michelle delved deeper into the complexities of nervous systems and how they evolved over the past half billion years. She compared this evolution to the much shorter time span over which computers had evolved from Turing's mechanical computing engine that decoded Germany's Enigma encryption device, to the first electronic computers using vacuum tubes as switches, and finally to the transistors still at the heart of all modern computers. She pointed out that both evolutionary pathways involved ever increasing numbers of neurons in the brain, and transistors in the CPU of the laptop she was using to project her lecture. She also noted that it was not just the number of units, but also the number of connections that were important. A transistor had just two connections for input and output, but a neuron in the brain had thousands of inputs called synapses, and a single output that branched to connect to hundreds of other cells.

Michelle went on to describe how she dealt with differences between animals and how she calculated the number of synapses transmitting information between neurons in the brain. She summarized the results in a table with brain weight in one column and the calculation of cortical neurons and their synaptic connections expressed as logarithms in adjoining columns. The animals ranged from humans with 11.5 billion neurons and elephants a close second with 11 billion, then on down through chimps and dolphins to rats with 15 million and mice with 4 million neurons. She also pointed out that when the logarithm of the neuron number was multiplied by the logarithm of the number of synapses per cell, and the product adjusted for the encephalization quotient, a new number emerged that was related to the complexity of the nervous system.

"Now, I won't bore you with the mathematical details, which you can find in my publications on this subject if you wish. For our purposes today, I will just show you the results in my last slide, then discuss the implications."

RELATIVE COMPLEXITY OF MAMMALIAN BRAINS
Human 45.5, Elephant 45, Chimpanzee 44.1, Dolphin 43.6, Gorilla 43.2, Dog 37.8, Rhesus monkey 37.4, Cat 32.7, Opossum 31.8, Mouse 23.4

"You might look at those numbers and ask yourself, 'Are humans only twice as smart as mice?' You must keep in mind that the numbers are produced by multiplying two logarithms, so the difference in complexity between the human brain and a mouse is immense, many orders of magnitude. For you dog lovers, yes, dog brains are more complex than those of cats. And if you think that dolphins approach human intelligence, well, they come closer than any other animal."

"If we asked a hundred thoughtful people like yourselves, and Dr. Akenov, to rank this list of mammals according to their own experience and observations, I predict that the lists, when the outliers are eliminated and results averaged, would closely reflect the calculated ranking. It is interesting that all five animals with normalized complexity values of 40 and above are self-aware according to the mirror test. The rhesus monkey is borderline at 37.4, and fails the mirror test, while the animals with complexity values of 35 and below are clearly not self-aware in terms of the behavior we can observe. I propose that the jump between 37.4 and 40 appears to reflect a threshold related to self-awareness."

"Although mammals with normalized complexity values between 40 and 43.2 are self-aware and are perhaps conscious in a limited capacity, they do not exhibit what we recognize as human intelligence. It seems that a normalized complexity value of 45.5 is required for human consciousness and intelligence, that is, 10–20 billion neurons, each on average with 30,000 connections to other neurons, and an EQ of 7.6. Only the human brain has achieved this threshold."

Michelle felt the relief of someone who has carried a burden to the top of the first and longest flight of stairs. She wanted to rest but she felt the audience waiting. "We now have a measure that might satisfy Lord Kelvin." She played her comic relief gambit and flashed on the screen above her the photo-shopped face of Lord Kelvin with a wide smile. The smile, however quickly faded to a frown. "Ah," Michelle said, "Lord Kelvin wants to know if we can prove our hypothesis. After all, he learned the hard way with his own miscalculation that the earth was only 30 million years old."

"We validate the hypothesis about consciousness and the complexities in nervous systems by proposing testable predictions. For instance, we might predict that in diseases such as Alzheimer's, the reduced intellectual capacity and lowered state of consciousness begin to occur when the number of active neurons or the number of synaptic connections is reduced below threshold values required for intelligent behavior. In fact, in patients with advanced Alzheimer's disease the number of synapses per neuron is reduced by 25–35 %. Similarly, when general anesthesia produces an unconscious state, we will find that the threshold is again breached, not in terms of cell number but instead because the action of the anesthetic compound reduces the number of functioning synapses and associated membrane receptors.

"From what I have told you, it seems reasonable to believe that it is possible to measure consciousness and intelligence in terms of complexity of the brain's connectedness. We can now understand what happened 200,000 years ago when the first humans separated from our primate ancestors. If we could ever reconstruct their genome, as we have done with Neanderthal DNA samples, I predict that we will discover a modern version of FOXP2, the human gene that separates us from other primates and gives us the capacity for language."

Michelle paused, a dramatic pause that demanded the full attention of the audience, then finished. "Finally, if there is ever another quantitative advance in human evolution, it will involve an increase in complexity of brain architecture, specifically the degree of connectedness represented by the number of synaptic junctions between cerebral neurons. Thank you for your attention, and I will be happy to answer questions."

She had to wait what seemed a long time for the applause to fade. The tall young Kazakh women with the feathers rising from the tops of their hats and their bare thighs and knee length boots stood ready to pass portable microphones. Michelle spent another half hour fielding questions. At the naïve level came questions on how to count brain cells and can we multiply synapses with drugs. Someone wanted to know if it would be possible to create a dog or a sheep with human level intelligence by multiplying brain cells and synapses. A young man followed by asking if it would be ethical. Michelle said she had no special expertise in ethics or philosophy, but that this gathering and this kind of question were recognition that humankind's progress in science was giving it powers once thought to be reserved for the gods.

When Akenov took the stage to thank Michelle and dismiss the session, he concluded, "As Dr. Murphy said, science has indeed given us god-like powers. I hope the sessions today and tomorrow can give all of us new insights into the nature of these powers and all the good that we can do with them for the world."

Chapter 16

At the Sunday luncheon that ended the conference Michelle had dozens of "vizitkas" which she thought was a nicer name than the English "business card". Avalon had enjoyed a very full program for the handful of teenagers who had attended, most of them Kazakhs with one or both parents at the conference. Akenov said to her, "Miss Avalon, you must have studied our language before you came. I congratulate you and we thank you for honoring our country."

Avalon replied, "The sponsors of the youth program and the other kids were wonderful teachers."

"You have maybe natural talent for Kazakh language," Akenov replied. "And I sure sponsors had exceptionally good learner."

Behind her polite smile Michelle thought, "*Avalon is probably learning Kazakh better than he has learned English.*"

After they had slid into their beds and turned off the lights, Michelle asked Avalon, "How was the youth program?"

"It was for youth," Avalon said.

"And for you?"

"Mom," Avalon said with an indulgent tone, "The other kids liked it. They had fun. But, Mom, I'm not a kid, not the way they are. I'm a mutant!"

Michelle lay quietly staring into the darkness. After a while she reached out across the space between the two beds, found Avalon's cheek and held her fingers there. She said, "Yes, it's possible."

As Michelle started to withdraw her hand, Avalon held it. For a moment their two hands clasped joined them across a divide greater than the darkness between their beds. Avalon said, "Thank you. Nobody has a mom like you."

Early the next morning Arman's driver picked them up at the hotel in a Mercedes SUV and took them to the airport for the short flight to Almaty. Instead of stopping at the terminal, the car drove out among the planes to a brilliant white Gulfstream where Akenov's small figure stood in the door waving an enthusiastic welcome to them. When he stepped aside to let them enter, he said, "President is in India and Singapore this week, so Ministry has done us special favor. We have his plane."

"Why doesn't he need his plane?" Avalon asked.

Akenov looked surprised but pleased. "This one belonged to the late President Nazarbayev. Mr. Ducinbekov likes a bigger plane."

He led them through the spotless, soft beige forward cabin into the slightly larger aft cabin and said that phrase that almost every Soviet schoolchild learned in the first week of English, "Sit down please."

Avalon and Michelle settled into the deep leather seats dyed Kazakh blue, facing forward, and Akenov took the backward facing seat opposite a small counter with a rolled down door over its interior shelves holding snacks and drinks. Michelle saw two large men in black suits come in and sit in the front cabin.

Akenov turned following her eyes. "Presidential apparat," he explained. "We can pretend I am president," he said.

Michelle had the sense that he might believe he could be president. "You are quite famous here," Michelle said. She meant it as bait.

"I am," Akenov said. "But Kazakhstan is a democracy. The people will decide."

Michelle had watched more than enough candidate interviews to note that he had not denied the ambition. She would put this in her report.

As soon as they had reached cruising altitude, yet another tall young Kazakh woman came back wearing a smile that Michelle had come to see as part of their uniform. She opened the snack bar and asked in perfect English, "Would you like coffee, tea, juice, milk? And we have cakes and cookies and fruit."

When Avalon said in Kazakh, "Thank you, I would like kumiz," the smile flickered off for only an instant. "The hostess replied in English that unfortunately they did not have kumiz."

Akenov laughed and said, "All kumiz is on ground, not in air."

As they approached Almaty, Akenov said, "There, on green slopes just below mountain forest is my home. Where I grew up. Unfortunately city is not very visible. It is down there in gray cloud. Too many cars."

Akenov had his face turned toward the window, and Michelle saw that no one watching her. She quickly exchanged her napkin for the one Akenov had used to wipe his mouth. When Akenov turned back to the table, she held up the napkin he would assume was hers and asked, "Would anyone arrest me if I took this home as a souvenir?"

Akenov looked at the blue napkin with the national seal in gold and said, "Small gift. In the president's absence, I say it is yours."

Michelle thanked him. Lipkovich had not asked for it, but why not include a sample of Akenov's DNA with her report? She folded the napkin very carefully and put it into her bag.

As they began their descent, Michelle once again caught a glimpse of the magnificent wall of snow-covered mountain peaks to the south that separated Kazakhstan from Kyrgyzstan. She remembered the difficulties she and Hank had faced in the city, and how the mountains had made their troubles seem small, renewing her faith that they had adopted an infant with that difficult name, Meruert, that the three of them were starting a wonderful new life. Bringing her thoughts back from the past to the present, she recalled holding hands with Hank sitting next to her as he had on their first approach to Almaty, only instead of Hank, now it was Avalon.

Avalon turned to Akenov and said, "Dr. Akenov, Do you know where I was born?"

Akenov raised his eyebrows in surprise. "Certainly not far."

"I wish I could remember," Avalon said. "Can you find out? I would like to see where."

Akenov said very firmly. "May be possible. I will determine." Akenov stared thoughtfully at her for a moment, then replied, "I can remember being born."

Michelle remembered the orphanage too well—the drab mustard colored linoleum floored corridors, the desperate look in the eyes of the older children. The matrons shouting orders, pushing curious children back into dormitories and play rooms. The clamor of the dining room. The air thick with grease and boiled cabbage with the occasional odor of sour milk. The aluminum spoons and the forks with only three or even just two of the tines left. The windows almost opaque with grime that closed the children into hopelessness. When she and Hank had first seen the 11 month old child whose name their translator wrote in English and tried to help them pronounce, the infant had made no sound and hardly moved, but her eyes fixed on Michelle's, then Hanks. They had not chosen her, she had chosen them. Hank had said it, "She wants you, Michelle."

"And I want her," Michelle had said with a certainty that surprised her.

Michelle had been glad that Avalon remembered nothing of that moment or the orphanage. Now that they were in Kazakhstan again and Avalon had become a self-confident young woman, Michelle felt differently. Avalon was entitled to know her full life—if it were knowable.

They spent the next day in the small town near Almaty where Akenov's Institute and fertility clinic occupied the former nuclear physics laboratories. Fields and small groves of trees surrounded the town of perhaps a thousand people. Some lived in the single family homes called "private homes" in Russian, and still the possession of privileged scientists or their widows. Most of the population lived in the two to five story apartment buildings on quiet streets lined with poplars. Michelle told Avalon about the earthquake in Almaty when she and Hank had come to adopt her. "I'd much rather be in a village like this," she said.

For a few hours Avalon accompanied Michelle and Akenov as they visited laboratories. Michelle looked carefully at the people and the equipment in the labs, checking for anything that could be related to biological weapons and the viruses that might interest Lipkovich. Nothing caught her attention, and she realized that the equipment for manipulating viruses would be universal, whether for ordinary research or fabricating a weapon. As they left, Akenov said in passing, "Those buildings next door are school for talented children—like you, Miss Avalon."

Avalon said, "I'd like to meet them." And while she doubted they would be like her, she did think for a moment that maybe what was different about her was something to do with being Kazakh.

Akenov replied, "We'll see. Maybe we can arrange for you to visit school."

In the early afternoon they drove to Akenov's house for tea. The house was a modernized two story home at the end of a cul de sac. Tall and densely foliated firs lined the front walk of red tile pavers. The house sat on almost an acre of land surrounded by a high iron fence with decorative but sharp spear points on top. A man in the now familiar black leather jacket opened tall iron gates to let the car in. At his side stood a large tan dog with deep fur and sharp clipped ears. He had stout legs and a thick chest. Avalon asked what kind of dog it was. "Shepherd dog. Kazakh shepherd dog who eat wolf," Akenov said.

Inside, both floors and walls were a gray polished granite, the stairs white marble, but the stony surfaces were softened by Kazakh felt wall hangings and woven carpets with loud greens and oranges but soft maroons and blues too. An elderly Kazakh couple came into the dining room to greet them. Akenov introduced his parents who spoke no English and little Russian. They beamed and bowed when Avalon greeted them in Kazakh. Akenov's father had a round leather face punctuated by bright small dark eyes and deep wrinkles. He was slightly stooped but broad shouldered and sturdy. Akenov's mother was a thin, small woman whose face was as weathered as his father's. She was not quite as short as her son. Despite her stature and bony hands, Michelle sensed the strength of rawhide inside and out. She wondered if she should describe them for Grable and Lipkovich. *Why not?* This part about being a spy she had begun to like. It was not so different than the first observations a scientist makes to test a hypothesis.

"Please, sit down," Akenov said sweeping his hand toward a large round table that sat only a foot off the floor, surrounded by quilted cushions. "It is table from parents' yurt." It was also a table at which Arman's short stature was less apparent. They sat cross legged at the table, the old man beaming, the old woman shuttling in and out of the kitchen bringing water, tea pot, the traditional Kazakh cups, little bowls without handles, and a large pile of puffed fried pastry call *baursak*. In the Kazakh custom, she kept their cups full of milky tea, careful not to let them drink to the bottom. At the old woman's first effort to refill, Avalon said, "I still have some," and picked up her cup, but the old woman took it from her hand, spilled the remains into a pot, and refilled it with fresh tea and milk.

Akenov explained with his strange smile, "If a Kazakh woman allows guests to empty their cups, it means guests should leave. We certainly don't want you to leave."

When they had insisted that they had drank and eaten enough, Akenov passed his palms in front of his face and chest in the gesture of thanks, and rose. Outside while they were looking at the livestock, Avalon was happy to accept Akenov's suggestion that she go riding with a young woman who trained the horses. "Our special breed," Akenov said. "Very smart, very gentle."

Avalon said, "I always dreamed of riding a horse." She paused, then quoted, "Oh, a wonderful horse is the Fly-Away Horse, Perhaps you have seen him before; Perhaps, while you slept, his shadow has swept, Through the moonlight that floats on the floor." Akenov looked at her with obvious interest.

"That is poem. You make it?"

"No, Eugene Field," Avalon said. "Mom used to read it to me."

Michelle smiled. "Go for a ride but don't fly away."

Avalon and Michelle slept soundly in the Institute's guest suite. Akenov had shown them in saying, "I sure that many future Nobel Prize winners sleep here. Maybe you, Michelle."

In the morning Michelle found a cream colored, greeting card-sized envelope that had been slipped under their door. From inside she drew out a single page of matching stationery with a hand written note. The ends of the letters curled elegantly. "Michelle, my driver Gani will arrive at nine to take you and

Miss Avalon to visit orphanage, then wherever you like. Apology I cannot be with you."

Just before nine a young woman came and escorted them to the street where a huge Kazakh with an obliging smile waited like a genie summoned from a magic jug. Without introduction, the escort said, "Gani knows the city very well. Dr. Akenov gave him suggestions, but if you want to see anything special, go shopping, have banya—just ask. He will take you to airport tonight. Your bags safe with him and you too. I have arranged special visit for you. This afternoon children's home where was Avalon will receive you."

Michelle's instant reaction was a protest which never reached her lips. She remembered the orphanage too well. And she had had enough of Kazakhs planning her life, however well intended. Avalon spoke first. "Wonderful," she said. "I want to know as much as I can."

Michelle drew a deep breath, looked at her happy daughter and thought, *she is old enough now to possess as much of her past as she can find.*

When they said they would like to drive into the mountains beyond the city, Gani took them up Dostik Avenue naming a few places as the broad street climbed gradually, then more steeply away from the city. The air grew cooler and fresher, and the constant smog of downtown disappeared. They declined Gani's suggestion that they try the shaslik, hunks of marinated mutton roasted over charcoal on a portable metal grill. They looked into the huge stadium cradled between mountain slopes, and climbed the hundreds of steps toward the dam that had been built to keep landslides from roaring as they once had down miles of slope into the city. Gani stayed protectively by their side as they walked up a trail into the tall evergreens. They rode the ski lift up to 12,000 feet where snow covered the bare rock. Avalon sat in the lift beside her mother, Gani in the chair behind theirs. Avalon said, "It's like a giant genie is taking us on a tour of the land of my ancestors."

"If you had grown up here, with your intelligence," Michelle said, "you might have changed the whole country—if they would let a woman do that."

"Maybe I'll return one day."

After they had stood a while at the highest station looking up at the last few hundred feet of mountain and down at the distant gray smudge of the air pollution blanketing the city. Gani said, "Soon we should be at the children's home."

"I'm ready, any time," Avalon said.

"I'm not, not really," Michelle said quietly. "I doubt we will see happy healthy children."

Avalon kissed her on the cheek. "But it's where you found me. I want to know what it was like."

The orphanage was little changed from how Michelle remembered it, maybe the paint on the walls fresher, the smells not as strong, but still the unpleasant aromas of cabbage and greasy meat. A woman in a bright white smock gave them a genuine welcome, then led them into several clean rooms strewn with toys and children playing, then a baby nursery with several nurses busy caring for infants. At one point Avalon began to stray from the rooms their host was showing them, and their guide almost panicked, saying almost sternly to Avalon, "Stay with me please."

"But I want to see everything," Avalon said veering off toward a dark and narrow corridor.

"Please! This way, I must insist."

Michelle realized immediately that this tour had been pre-arranged, sanitized. Avalon smiled and asked, "Are you hiding something?"

The guide said only, "You must stay with me or you must leave."

"But this was my home," Avalon said, still very pleasantly. "Did anyone here know me as a baby?" she asked.

"Almost everyone is new since then," the guide said. She quickened her pace toward the director's office.

Michelle recognized the director, but she was not sure the director recognized her. Their meeting lasted barely beyond the introduction, long enough for the director to thank them for coming and to assure them that care of the children has greatly improved.

Afterwards, outside, Michelle said to Avalon, "Well, did you remember anything?"

Avalon laughed. "I remembered that I didn't like being here. And you know, I feel sorry for all those little kids. It felt like looking through a mirror into the past."

Chapter 17

"Meruert Sagindikova." The catastrophe at the Almaty airport began when Michelle heard the two words spoken. Twenty-seven hours later when Michelle finally lay down on her bed in Eugene and closed her eyes, she kept hearing the name and thinking that if she could banish that name, she might have Avalon at her side. She had not slept on the flight from Almaty to Amsterdam or in the Schipol airport or on the flight to Portland or the half hour hop south to Eugene. And sleep still did not come to take her out of the twisted dream she lived in. As she and Avalon had sat relaxed in the airport lounge in Almaty, four uniformed officials, three middle aged men and a very severe fiftyish woman, approached. The woman stopped in front of Avalon and said the name *Meruert Sagindikova*. Avalon had been born with that name.

In a small windowless office decorated by a more than life-size official portrait of the president looking calm and fatherly, the three men stood along one wall, and the woman sat behind a desk where her thin folder of documents was the only object on its worn surface. Michelle and Avalon sat in folding chairs facing her. The woman opened the folder, looked at the top document, then at Michelle. "This young woman is Meruert Sagindikova."

Michelle answered, "I don't know anyone named Sagindikova. She is my daughter Avalon Murphy, you have taken her passport and soon we will miss our flight."

"You will not miss your plane, Dr. Murphy," the woman said. "You must understand that her American passport is a passport, not her identification."

Michelle willed herself to be calm. "It's her American identification. It's who she is now—an American citizen. My daughter." Michelle emphasized the possessive. "And besides, what is this all about? What difference does it make who she was? She is legally my daughter."

The woman stared at Michelle, then at Avalon, then at the passport which she folded and slapped on the desk as if it were a playing card. "In Kazakhstan she is legally Meruert Sagindikova and she is legally a Kazakh citizen."

Michelle uttered only, "But ..." when the woman held up her hand. She showed no trace of sympathy or concern for Michelle or for Avalon. Michelle read that as a warning that the woman was about to deliver bad news. She looked over the woman's head at ex-President Nazarbayev's portrait, not yet replaced, his kindly smile beaming down at her.

"I will explain. After you entered the country we examined her records. The adoption had many irregularities. It was not complete."

Michelle remembered many irregularities—delayed papers, missing papers that appeared after payments "to facilitate research", hundreds of dollars in fees to notaries who could validate this or that document, and more hundreds for pastille seals on translations that no one had known were needed. She remained silent. The three men standing against the wall moved their eyes from Michelle to their woman colleague. Avalon sat straight and calm in her chair. When Michelle looked at her, she smiled a smile that said, *I'm fine, I'm listening.* Michelle felt as if Avalon were watching her play a game, encouraging her. Michelle again looked up at President Nazarbayev. This was his country.

"Under our law, Dr. Murphy, a child adopted by foreign nationals retains Kazakh citizenship until she turns 18. She cannot decide to give up Kazakh citizenship until she is 18. It is also the law that she may only enter and exit Kazakhstan using a Kazakh passport."

Michelle had decided her best response was silence, at least until she knew what game they were playing. What was the point? The three men looked from her silence to their colleague again.

The woman continued. "The passport might be issued in a few days, but unfortunately there are several other more serious irregularities."

Michelle looked up at President Nazarbayev, then asked with a very studied calm, "And how much would all this cost?"

The woman looked up to the three men and let out a deep breath, perhaps a sigh. "Dr. Murphy, bribery is a crime in Kazakhstan."

Michelle gritted her teeth. "I wouldn't think of offering a bribe," she said.

"But you did when you and your husband came here to adopt Meruert. In fact we know of several bribes."

Michelle glanced for the last time at President Nazarbayev, but his smile had become a smirk. She said silently, *I am not amused.*

During the following 27 hours Michelle replayed over and over the airport conversation and the many things she could have said. They had allowed her to call Arman Akenov and the US consulate. A Marine guard had answered the phone and said the consulate was closed. She insisted he call the consul. Very politely he said in a Louisiana accent, "M'am, I cannot do that." Should she say she was working for the State Department? With the CIA? Anyone could say that and no CIA agent would call from a cell phone. She left her name and cell phone number. Akenov said everything must be a huge mistake—"confusion" he had called it. He had Michelle pass the phone to the uniformed woman and they spoke for a minute in Kazakh before the woman returned the phone to Michelle. He told Michelle that these were dangerous people, and she should be very careful what she said. Recently papers had printed stories about children adopted by foreigners being abused, even sold as prostitutes. Public feeling about foreigners adopting had grown increasingly bitter and several candidates for the fall election were using it as an issue. "I'm beginning to feel bitter myself," Michelle said.

Akenov replied, "Yes, I understand," in a way that meant he didn't under-stand or didn't care. Michelle's mind began seeking connections. If she asked if he knew about this problem, of course, he would say no. She would bring it up with Lipkovich.

Michelle called the consulate again and begged the guard to call the consul, and if not the consul, then the ambassador in Astana. The guard said he would see what could be done. Michelle called Akenov again. Instead of asking if he were involved, she asked, "Can you think of any reason they would do this?"

Akenov confided in her that he had learned that Avalon's birth mother had been found. He thought the woman wanted money, but she would not say so. He talked to the uniformed woman again. When Michelle again took the phone, he told her that he had been assured that Avalon could be released into his custody until everything was resolved. "I cannot get to airport before flight. But be sure. No more than hour and Avalon will be with me. It is my promise."

Michelle considered. To know who Avalon was with and where was better than the unknown. And did she have a choice to deny Akenov? A CIA staged hostage rescue scene popped into her mind, and she brushed it aside as television stuff.

Michelle stood up, placed her palms on the edge of the woman's desk and insisted on staying with Avalon. The three men shifted to a more alert posture. The woman said, "We have agreed that your daughter can stay with Dr. Akenov. You are not satisfied? You have violated our laws coming here. You violated our laws when you came here to adopt a child. You are not in a position to object." She let this sink in. The three men seemed to hover at the ready.

The woman shifted to a conciliatory tone. "Meruert will be very safe with Dr. Akenov. He seems very concerned and will act on your behalf. He is an important person. Meruert will stay here, and these officers will take you to the plane before it is too late."

Michelle turned to Avalon, who said, "Mom, it's all right."

Michelle almost laughed at Avalon's confidence. A newly minted 13 year old telling her mother that she didn't mind being kidnapped. Michelle's phone rang.

"You must go now," the woman officer said firmly.

Michelle answered her phone. She turned to the woman and said the words she knew would win her time. "The consul." The voice on the other end was young. This was a minor consul on duty after hours, a junior post. After thanking the consul for the quick return, she explained the situation quickly. She had decided to be calm, to control the emotion in her voice, but this had no effect. The consul said he had no authority that would allow her to stay or force them to release Avalon. "In short," he summed up, "Their laws about the status and citizenship of adopted children are complex. They may not be fair, but they are correct if they say a foreign adoption has not complied with all the rules. Perhaps they are deliberately designed that way."

"I want you to contact someone I know at the Central Asian Desk in Washington," Michelle said. When she said Rhonda Grable's name, the consul said nothing for a few seconds.

"Of course, the embassy will talk with Washington," the consul said.

"When?"

"I will talk with the ambassador in the morning. I mean later in the morning."

Michelle looked at the time on her phone: almost 2 a.m. She had the sense not to say she would also talk with Grable in the morning.

The luxurious normality of the first class flight from Almaty to Amsterdam's Schipol Airport almost mocked her. At first she declined the glass of champagne, then changed her mind, hoping alcohol would shut down the worst in her mind. The slightly plump, pear shaped man next to her had taken his glass and even urged her to have one too. "I'm celebrating," he said, and she heard a faint southern accent.

Michelle replied silently, I'm not. As soon as she changed her mind about the drink, the man introduced himself, "John Leyens," he said. "From Dunellen Springs, Florida. I just got engaged." He handed her his business card.

John Leyens, CEO
Reconditioned Aviation Components, Inc.

Michelle read the card and raised her glass out of politeness and he brought his over to clink on hers, a plastic clink.

"You pay bribes in Kazakhstan?" Michelle asked.

"Bribes? I don't do business here. Maybe my fiancée will. She's Khazakh. But, yeah, I guess you could call them bribes, all the fees we had to pay notaries to do the papers so we could get married. Never heard of such expensive notaries." He laughed. He went on without stopping, telling Michelle how he had met his fiancée, Natasha, by Internet, talked to her on the phone, agreed to get engaged and apply for a fiancée visa. "I know," he said, "I'm not very attractive to women, okay, but I'm not dumb—she's single, she's Russian, she wants to get out of Kazakhstan, and maybe I'm just her ticket. So, worst case—I've helped her escape. I haven't helped too many people—I mean not personally."

Michelle turned off her overhead light and said, "Sorry, but I need to get some sleep."

"Wake you for breakfast?" he asked.

"Please don't," she said. She turned away and pretended to sleep.

During the flights home Michelle many times had thought of possible ways she could have stayed with Avalon or at least close to her. When Don and Jodie picked her up in the darkness of early Saturday morning in Eugene, Jodie immediately sensed that something was dreadfully wrong.

"Michelle, what happened? Where's Avalon?"

Michelle shook her head. "I can't talk about it here. Let's get my bags. I'll tell you later." At the baggage claim, when two suitcases appeared, hers and Avalon's, she just pointed. Don glanced over at Jodie, then pulled them off the carrousel and followed the two women out to their Accord waiting in short term parking. He unlocked the car and put the suitcases in the trunk while Jodie held the back door open for Michelle, then got in beside her. Neither Don nor Jodie said anything on the trip home, respecting Michelle's painful silence. Don was easing into the driveway of Michelle's condo when she finally said in a low voice, almost a whisper, "I had to leave Avalon in Kazakhstan. They said our adoption was illegal."

Over the next hour, the three sat in Michelle's kitchen having bagels and coffee while she told them about the conference, how quickly Avalon picked up basic Kazakh words, Akenov, the trip to the orphanage, and finally the nightmare at the airport.

"It was my last question to the consul who made me get on the plane. I asked him if he thought they could take Avalon away from me—permanently. He said they can if they want to. It was like being cheated, robbed, and raped, only I don't have a single bruise to show for it. They forced me to abandon my own daughter. We're citizens of the most powerful country on earth, and petty officials in Kazakhstan walked away with my daughter."

She went on to explain, to their sympathy and disbelief, everything she had learned about the adoption laws of Kazakhstan and how the authorities there could prove she had broken their laws. The formal consent of Avalon's biological parents had not been obtained—no matter that the orphanage managers had said they did not know the parents. There had been no formal court finding of the "moral and personal qualities" of Michelle and Hank prior to adoption. Because legally Avalon was a Kazakh citizen until 18, Michelle should have filed an annual report "on health and living conditions." Finally, Kazakh officials said they had proof that the adoption had used bribes. "I almost asked them how much I needed to bribe them," Michelle said. "I did ask them whom they were working for—why they had chosen Avalon. You know what they said?"

Don and Jodie asked with their eyes, knowing the answer was coming.

"One of them said, 'We work for Meruert Sagindikova who is a Kazakh citizen.' Another said, 'And for the people and government of Kazakhstan where she is a citizen.' And they said it with straight faces."

Michelle's own face crumbled and she hid in her cupped hands.

Don and Jodie asked if she wanted them to stay with her as she settled back in. "No," she laughed bitterly, looking up. "Thank you, but I'm in no danger. Here I just feel so helpless. I might not even have gone to the conference except . . ." No, she would not bring up Grable and Lipkovich and the CIA. And besides, she would have gone anyway, and she would have taken Avalon. She saw Jodie and Don look at each other simultaneously. "Really, I'm all right, and I think I can sleep."

"Call us?" Jodie said.

"I will. I think I'll need your help."

"Just ask," Don said, standing and stretching.

As each gave her a hug at the door, Don said, "You're not helpless here."

Michelle felt at home in body only. Her thoughts, her being, her daughter were in Kazakhstan. She should still be there. She had not fought hard enough. As she lay on her bed she opened her eyes in the darkness. She began a series of supposing that had no answers. Had Akenov done everything he could do? He was a scientist, a colleague, an odd man but important in his own country. How could she use him? It was going to be a nice fall day, and as the morning sun began to reveal the familiar objects in her bedroom she considered what she would do. She also realized the time in Washington was three hours later than in Oregon. She had to call Grable and Lipkovich. No, first Lipkovich then Grable.

Chapter 18

Don and Jodie reserved their weekends for each other. Even if they did not go out to the coast or the mountains, they kept their pact—no radio, television or newspapers. Often they did not answer the phone, but this Saturday they were waiting for Michelle to call. The phone rang at 11, and minutes later they were out and walking briskly to Michelle's. When they arrived, she thanked them for coming quickly and began to apologize. "I know your weekends are sacred. . ."

Jodie said, "Stop. Please stop." And when Michelle did not go on, Jodie hugged her and whispered something in her ear, while giving Don a look that meant, stay here and wait. With an arm around Michelle's shoulders, Jodie guided her down the hall as if Michelle were in a strange house. Don watched them go. He mused, *Homo sapiens at our best.* The two women disappeared down the hall into Michelle's bedroom and the door closed, closing Don out, but he knew Jodie had chosen the right first step.

Don looked at a picture above the mantle—Michelle with the infant Avalon tucked inside a thick winter coat, her tiny face just visible above Michelle's lapels and under a thick felt cap with a feather on top. Hank was smiling down on her, and she was smiling up with her dark eyes in her brown face saying, "Yes, I'm glad I'm yours." *Well, something like that.* Behind them rose the grand white wall of the Almaty Range of the Tien Shan. So much beauty, great and small in one picture, Don thought. Then he imagined the picture without Hank, then without Hank and Avalon. *It just can't be that way*, he thought.

He listened for any sound from the women. No sound was good he thought, but he didn't know how he knew that. Maybe from his own blissful silences with Jodie. Sometimes he thought of them as *harmonic silences*. He shared that kind of silence with only one other person—Avalon. She had that peculiar ability to be both occupied with her own thoughts and present for a friend at the same time. He wondered now and then if this ability somehow was the inward reflection of her ability to move her left eye independently of her right eye, something she seldom did in company. When Don had expressed a scientific interest in that ability, she demonstrated without hesitation. She had not just a quiet confidence, but he'd thought of it as a superior confidence though he wasn't sure what that implied. He felt it every time he explained something to her or taught her how to do something. She was never confused. She followed as matter-of-factly as if she were the teacher and he was a student answering a test question. He never had to repeat even part of an explanation. Physical skills—doing a card trick or hitting a softball—learning these she did no better than her peers, sometimes worse. Anything she couldn't understand on her first try she analyzed, breaking it down into smaller segments. Don knew about the old argument between Hank and Michelle, and he was on Hank's side. He thought both Michelle and Avalon were now ready to find out what made Avalon different.

Don looked at more pictures of Avalon playing, and a couple of pictures of Avalon and Michelle visiting the dig in Africa four years ago. He didn't know how long he had been alone when he heard the bedroom door open. He said nothing while they sat together on the sofa opposite his chair. When Michelle leaned forward slightly, he realized he expected Jodie to start.

"Michelle thinks there is something we can do," Jodie said, turning to Michelle.

Michelle said, "I'm pretty tough, I think—'fighting Irish' and all that, actually a family of fighting Irish alcoholics—a long sad way from Tipperary. But I can't fight this one by myself, and ..." Again she hesitated. She said, "What I can tell you is that I'm not alone in this."

"You mean someone else besides Avalon?" Don asked.

"I'd like to tell you the full story," Michelle said, "and I will if you want it, but first I have to say two things. I promised not to tell the whole story, and you may not want to get involved."

"You know the answer to that last part," Don said. "If you want us, we're with you."

Jodie asked in a tone of delicate concern, "Who did you promise and why?"

Michelle took a very deep breath, let it out and said, "The CIA."

For several long seconds there was silence while Jodie and Don absorbed this surprising statement. Having no doubt that she could trust them, Michelle began at the beginning. "I should have told you when they first came to me."

Half an hour later, Michelle ended the story with her conversation early that morning with Evan Lipkovich, who had already been in contact with Rhonda Grable when the embassy in Almaty called her. After Michelle told her story, Evan had said, "I'm so sorry. We didn't expect anything like this. There's no way that Akenov could have known."

"I hope sorry isn't where it ends."

"I understand," Evan replied. "I'll be there as soon as I can, probably tomorrow morning. I want to hear everything in as much detail as you can remember."

"I remember very vividly," Michelle said. "I have a lot of notes, but I haven't organized them into a report."

In his office in Washington, Evan Lipkovich did not know the right words. He knew the wrong words, and he knew trite words, and they were also the wrong words. Silence could also go wrong. "Are you ready to go back to Kazakhstan?"

"I am, but let's forget about Akenov until I get my daughter back."

"I understand," Evan said. "By the way, have you had any contacts with Kazakh nationals here in the States?"

"Of course. That's why my husband and I decided to go to Kazakhstan."

After a few seconds, Evan said, "Tell me about them."

"Well, when Hank and I were at the University of Indiana, we met a wonderful young woman named Botagaz. She had also been in an orphanage as a child, and told us that Kazakhstan had many orphans available for adoption."

"I see. Anyone else?"

"Yes, we have another Kazakh student here in Eugene. His name is Bakhyt, but I only met him once. He has a scholarship to support his studies and said he was interviewed by Akenov as part of the process."

"I'd like to meet him. In private. Can you arrange that?"

"I'll call right away and make sure he's around."

"Good. See you tomorrow."

She almost called him back. Did he think Akenov had anything to do with Avalon's detention? She would wait for a few more hours. Then she remembered the napkin which might have Akenov's DNA. Somehow she wished she could use that to get Avalon back. Maybe his DNA would tell them if he was an autistic savant of some kind.

Don and Jodie were silent for a while when Michelle finished. Michelle said, "Don, I'd like you to be with me when Evan Lipkovich comes. I don't like secrets, and they are not going to shoot me. I'm entitled to the support of a friend and colleague, and they'll have to live with it. Avalon's what's really important," Michelle said. "I'm going to get her back."

Neither Don nor Jodie offered the false reassurances that everything would be okay.

"You know we're with you," Jodie said.

"But where are we?" Michelle asked. "The more I think about it, the more I'm convinced that it's much more than bureaucracy or even corruption."

Jodie saw Michelle's eyes move to the photo of her and Hank and Avalon and rest there a moment, then come back to hers.

"Damn," Michelle said. "I know there's more, but I don't see it. I feel like I've gone blind."

"We're all always part blind," Don said. "What's the line from Thoreau—it's not what we look at but what we see that matters?"

Jodie reached a hand toward his face and stroked his cheek and gave him a smile only he could recognize. It meant, we agree, let's move on.

"Let's think for a moment—do the facts suggest anything? Anything at all? Maybe we can at least form a hypothesis. Michelle, you were going to say something earlier."

When Michelle shook her head, Don continued. "Who and why?"

"I can think of why," Michelle said. "Who *wouldn't* want Avalon?"

"Okay," Don said. "Let's think about who. No holds barred, nothing's too far out if it fits." Don had been pacing, but now he sat down. He began to think as he did when confronting a new scientific question. He closed his eyes, sat up as straight as he could, let the tension drain from his shoulders, and folded his hands in his lap just below his navel—his version of a meditation position. Instead of emptying his mind, however, he summoned all the known facts, and to see them in the mind's eye as a list printed in white against a black void of the unknown. For a minute or two they sat in silence, asking from their imaginations to come up with possibilities that would not be pure fantasy.

"Mother who lost her child," Jodie said.

Don flinched inwardly. He and Jodie had wanted a child. Two miscarriages, but they continued to hope.

Don added another possibility. "Somebody in the Kazakh adoption business who will ask for more money."

Jodie and Michelle looked at him with small frowns of doubt, though they knew this brainstorming game very well and would not shoot down anything. Brainstorming didn't work when people went negative.

"Birth mother," Michelle said. Don and Jodie glanced at each other with 'that's possible' looks.

Don added, "Birth father. Brothers or sisters, uncles or aunts. Kazakh families are close."

Their list grew, drifting toward the less and less likely and even absurd—the orphanage as a child buyer and seller, someone from the conference, someone who had always wanted this child and had been waiting for her to grow up. They even speculated on a CIA plan to create an international incident that would serve a purpose they admitted they could not imagine.

Michelle added, "I read that in traditional Kazakh families if an older brother dies his younger brother marries the widow and becomes father." Michelle thought to herself, *Are there really so many people who would do this? Was I naïve? Is the world such a dangerous place?*

Don was thinking, *Among the variations in the human animal, there are some very bad ones.* He was embarrassed to find his mind wandering to the early twentieth century eugenics movement in which people like Margaret Sanger were moved by fear that the bad variants would take over. Almost before he knew why, he blurted out, "A Kazakh Hitler."

Jodie and Michelle both stared at him. "What I mean," he said, "is like what Hitler wanted—blond, blue eyed Germans, only he focused on eliminating everyone else."

Jodie spilled out three more possibilities. "Child prostitution—sorry, but possible; political power play of some kind; mistaken identity."

Michelle said silently, *No mistake.* But she showed no reaction and played the game.

When the silence grew long, Don said, "Let's look at the list, hard, just think for a few moments." In a while, he said, "Let's each write down the three we think most likely, number 1 being at top choice." A moment later he said, "Let's compare."

They put their slips of paper face up on the table. Both Michelle and Jodie had birth mother at the top of their list. It was second on Don's list, with "bribe seeker" first. Michelle's third choice was Kazakh Hitler.

Michelle gave a bemused sigh. "We're pretty conventional thinkers. I put birth mother only because the other choices seemed so far out. But her birth mother is obviously a Kazakh in Kazakhstan."

Don started to say something, hesitated, then said it. "Reasonable to assume, but let's keep all possibilities open."

"Yes, of course," Michelle said.

Jodie asked, "Do you know who her mother or father were? Or maybe are, present tense?"

Michelle sat up straighter and leaned forward. "*They* may have been birth parents, but I am her mother," she said, and immediately added, "Of course, we changed her first name from Meruert to Avalon, and at the airport the woman used the family name: Sagindikova. I suppose their full names are something for which we should have paid a few more bribes. None of the adoption people or orphanage people would say anything more than her Kazakh first name and that she came to the orphanage from a clinic in Narynkoll. That's east of Almaty almost on the border with China."

"So someone knew her name and where she came from and who brought her," Jodie paused, then added, "Or bought her. And someone at the clinic probably knows at least who her birth mother was."

Don sat back, listening and taking meditative sips from his coffee. He was sure the two women would pursue this better than he could. In some way, it might even be good for Jodie, a kind of odd way of getting a child or at least a way of knowing there's really a child out there. He hoped. For a moment he lost track of what the women were saying. His mind wandered to the question of how evolution could have produced an animal that ends up in a room full of furniture made by its own species, contemplating the kidnapping of one of its children. Kidnapping, of course, was not exclusive to humans. Which led him to the thought, *Would humans ever evolve to a condition where they were not recognizable as animals? Or,* he thought, *will we engineer that evolution?*

Jodie broke the silence. "We could make a list of everyone who knows Avalon here in Eugene. Maybe there's a local connection, because otherwise how could someone know that Michelle was traveling to Kazakhstan?"

"Who couldn't know?" Michelle asked. "The program was on the Web, in the University's PR releases."

"Long list," Don said. "But let's take a few minutes to think about someone who actually *could* do this."

Michelle got up and brought in the coffee pot and refilled the mugs. They thought. They came up with one name—Bakhyt. He had known that they were going to Astana, and Bakhyt had known Dr. Akenov. Don said he would talk to Bakhyt.

"Before you do that," Michelle said, "Evan Lipkovich also wants to meet him."

More silence passed, then Michelle got up and began pacing. She was a good runner. She wanted to be running after someone, the kidnapper, running him down. But she didn't know what direction she would run in. She stopped and faced Jodie and Don and pumped her fist in the air the way she did when emphasizing a point. "Akenov! Akenov knew everything about my trip. Akenov got us to Almaty. Akenov knew when our plane would leave. Akenov arranged for us to fly out from the VIP lounge. Akenov suggested he could take care of her. Shit! Why didn't I see it!"

Chapter 19

Almaty, Good Shepherd School

The warm sunlight of late afternoon flooded through the corner windows of Dr. Arman Akenov's office, falling on two leather upholstered chairs and a couch arranged around a low table, a slab of dark brown rapakivi granite from Finland. In the middle of the slab, almost like a Christmas crèche, stood a finely detailed miniature bronze sculpture depicting a flock of sheep with a shepherd sitting tall and vigilant in the saddle of his horse. A single piece of folded paper rested next to it with a lovely golden sculpture of a running horse resting on it as a paperweight.

The scientist sat quietly at his desk, also a granite slab, reading once again a copy of Michelle Murphy's paper about the unusual DNA sequence she had discovered in her daughter's genome. A hand knocked softly on the door just before the door opened wide enough for his secretary to look in.

"Dr. Akenov, Asel is here with the girl."

"Yes, send her in."

The door opened farther and Avalon walked cautiously through, followed by Asel Satpayev with her hand on Avalon's shoulder. Akenov stood and nodded to his secretary, who backed out and noiselessly closed the door.

His first test was simply to look at the girl, watching how she responded to his gaze. He saw a pretty girl, destined to be tall. He thought her face was a little long and thin, but the girl was also thin. She had good black Kazakh hair, but unusual brown eyes. She stared at his face for a moment, then her eyes began to move about the room, taking in the thick, hand woven carpet with its complex design, the large painting of Kazakh horsemen racing through a cloud of dust, the sunlit window with the Tien Shan range gleaming white in the distance. Her eyes returned to his face. She seemed unusually calm, considering that she had been abruptly taken from her mother by a group of strangers. Perhaps she liked being in Kazakhstan, he thought. It was her home, after all.

Akenov made his face smile. "Welcome, Avalon," he said in English.

Avalon stood perfectly still and quiet, eyes meeting his.

"Do you know why you are here?"

"I know what they said at the airport about my adoption."

Akenov was pleased. She was smart. She was strong. She was Kazakh. If strangers had taken him from his mother as a 12 year old, he would have behaved in precisely the same way. And he would be planning to win an advantage. He laughed, and motioned toward the furniture at the other end of the room. "Well, let's go sit down and I will explain. This is all temporary. In a perfect country with perfect people there would be no rules. For more than

70 years Russians made our laws, and we are trying to straighten out all that and live again like Kazakhs—like modern Kazakhs."

Akenov settled himself on the edge of the black leather couch and gestured toward one of the chairs. Avalon looked up at Asel, who smiled and with a very gentle press from behind guided her forward. Keeping her eyes on Akenov, Avalon slowly walked to the farthest chair, followed by Asel. She didn't sit, but stood in front of the chair and faced Akenov. Asel leaned toward her and whispered, "Don't be afraid. Dr. Akenov wants to help you."

Akenov caught her eye. "Asel, I think time for tea. And please bring some of that delicious black currant jam that you like so much."

"Yes, good idea," Asel said, then asked, "Avalon, do you like your tea white?"

Avalon looked up at Asel. "White tea?" Arman noticed that she had the typical high voice expected from a girl her age, but that her pronunciation was very precise.

"Oh. It means with milk."

"We don't drink tea at home."

Akenov said, "At home in Kazakhstan all people drink tea, even young children."

Asel said, "I think you'll like it white, with milk and a little sugar as we drink it. I'll be right back."

After Asel left, Avalon sat down. Although she was outwardly calm, Akenov noted she sat on the edge of the chair, as though ready to spring up and run.

He smiled again, and pointed at the folded paper on the table. "That is document from our immigration people. It allows you to stay with me until . . . until problems are solved."

He watched Avalon carefully as he said this. He saw her eyes widen, and a sharp intake of breath. He reached down and removed the paperweight sculpture, then slid the paper across the table so she could reach it. He saw Avalon look at the paperweight for a long moment, then she picked up and unfolded the sheet. She glanced at it but tossed it down immediately. "I can't read Kazakh, yet," she said.

"I just wanted to show you I have this document." Akenov made his face smile again. "So you see, Avalon, really nothing. There was bureaucratic nonsense about your adoption that needs to be taken care of. Sometimes people follow these laws. Sometimes they do not."

She had watched his face closely as he spoke and she saw his eyes shift momentarily to the left. When his eyes returned to lock on her face, they had that unnatural challenge of steady stare. The too steady stare of a liar.

"Why do they follow Kazakh laws for me? I'm American."

"American in Kazakhstan, and until you are 18, our law says you are our citizen. Maybe you understand—our people see beautiful Kazakh girl who is still Kazakh citizen. Maybe they want you to stay."

Now he wasn't lying. "And do *you* want me to stay?"

Akenov caught the question like an unexpectedly thrown object. Holding it in his mind, he bent his head down and lifted his tea cup for a contemplative drink. "Yes, I would like you to stay." Again, she could see his face relax, his palms turn up. He was telling the truth. "I am glad you will be with us, even for only short while. And I think your stay will be short. However, I'm afraid that in such cases our laws don't allow you to leave country until everything is in order."

"Do you have any children?" Avalon asked.

Akenov inwardly winced. Another unexpected question. Another sip of tea. He understood she was very smart. Maybe as smart, perhaps smarter than he was. He also knew he could still win what he wanted. One thing he wanted was the smartest Kazakhs he could find, and she was the smartest he had ever met, perhaps with one exception, and he smiled as he thought about that exception. He said, "I have many students but no children. And now, for at least few days you too are my student. I know your mother professionally, and I invited her to come to meeting, so it seemed only courteous for me to take care of you for few weeks here in Almaty. Better than strangers or children's home."

Avalon shrugged. "I don't know anyone here. There is no choice for me."

"You know me. Not well, but you will. I would like to ask you question."

Avalon said nothing, her posture signaling assent.

"Do you ever think of yourself as Kazakh?"

"I know my parents were Kazakh, but I am Avalon Murphy."

"Yes, you are—in America. But here we are all Kazakhs—you too, and Kazakhs are never strangers to each other."

"What is this place?" Avalon asked. "I saw kids playing outside, and it looks like a school."

Akenov relaxed. She knew she was in his possession.

"It is school, but I think in your country it would be called school for gifted children."

"Talented and gifted," Avalon said. "Special education we call it."

Akenov smiled. "Yes, each student here very intelligent, and each has at least one special talent." He paused for a moment. "I think you are gifted too. That's why my school will be good place to stay while you are waiting for your mother. Do you have special talent?"

Avalon shrugged modestly. Of course she had special talents, but the less this man knew, the more advantage she had. She stood and walked to the window where she could look down at the school yard. She heard the muffled mixture of shouts, screams and laughter that can be heard anywhere that school children gather to play.

Still peering out the window, Avalon asked, "Why do you think I'm gifted?"

Akenov leaned over and retrieved the note. He was amused and laughed. "I see that you think for yourself, which is what smart people do. But tell me, since you will be here for while, would you like to see how is our school? You could go to class and meet other children your age."

Avalon shrugged again. "Maybe."

"What school year were you in at home?"

"I am not in a year. I take special classes or any classes I ask for. And my mother teaches me."

Akenov was not surprised. "See, you are gifted as I said."

Akenov thought for a moment. "Avalon, would you mind some questions? Then I can see which classes you might enjoy here at Good Shepherd."

Avalon shrugged again. "What's Good Shepherd?"

"Of course, you wouldn't know that. Good Shepherd name of our school. But you just asked question, so now my turn. In your schooling, how far did you get in arithmetic?"

Avalon paused, examining Akenov, weighing whether to answer. Then she took a deep breath, reaching a decision. "I learned arithmetic when I was 3 years old."

"Then it was your mother or your father who decided to teach you. Why?"

"I asked them to teach me. I saw how they used numbers. How they counted. I liked the way that numbers described what I was seeing, how numbers fit together in interesting ways."

"Can give me example?"

"Some numbers fit into other numbers, and others stand by themselves."

"You mean prime numbers?"

Avalon nodded.

"What most interesting thing about prime numbers?"

"I wondered at first whether there was a pattern. Also, how many there were, if they were endless, and if not, which was the biggest."

"Those deep thoughts. Did you reach conclusion?"

Avalon shrugged. "When I learned to read, I found out that other people have been asking the same questions and had some good answers, so I stopped thinking about prime numbers."

"What do you think about now?"

"Well, everything. Mostly science stuff." In fact, she was being tutored in math by Professor Botvinnick, a brilliant topologist at the University of

Oregon. Don Koskin had been feeding her a series of anthropology books, and she had already absorbed the basics of genetics from Michelle.

"Here is question," Akenov said. "why does sun rise in morning?"

Avalon frowned. "It doesn't rise. It just looks like it does. Earth rotates and when it has turned us in the right position we have either day or night."

Akenov had never spoken with a child from a western country, and was surprised by Avalon's directness. Kazakh children could be intelligent, but they were taught to accept what they were told, not to think for themselves. Just then the door opened and Asel backed in with a tray carrying a tea pot, three cups, a basket of puffed pastry, and a small glass bowl filled with dark red jam. Kazakh abstract designs in deep blue with gilt outlines decorated the porcelain teapot and the cups without handles. Asel carefully set the tray down on the granite, then put a teaspoon of sugar into a cup she set in front of Avalon and filled it a third full with milk. She poured the tea from several inches above the cup so that the milk foamed. Asel turned her palm up by the pastry and said, "Our baursak. Tell us what you think of our tea."

Avalon sipped, then put the cup down next to the paperweight, pausing there just for a second to examine it. "The tea is still pretty hot, but I like it. Rakhmet."

Asel raised her eyebrows. "Dr. Akenov, I told you she was smart. Avalon begins to speak Kazakh."

Akenov looked at Avalon suspiciously. "How much Kazakh can you speak?"

"Words and phrases I heard at the conference on the translator earphones and from kids in the youth program."

When Asel had served Akenov, he said, "Avalon thought she might like to see our classes, learn Kazakh language."

Asel said, "I can arrange that with the rector. I believe Avalon will soon speak Kazakh."

"She is Kazakh," Akenov said.

Avalon was about to say again, *I am an American*. She said, "Yes, I want to learn Kazakh."

Asel sat down, ate a baursak with a small spoon of jam in the Russian tradition while Avalon sipped her tea. After a few minutes of silence, Asel glanced over at Akenov, who nodded. Asel finished the last of her tea, then said, "Avalon, when you are ready I'll take you to the dormitory and introduce you to the other children. Your suitcase is in your room, but if you need anything else just let me know."

"What about my classes?"

Akenov laughed. "I'm afraid they could not teach you anything even in our most advanced science and math classes. You can learn history of Kazakhs and

Kazakhstan. You know string theory but you do not know Kazakhstan. I am correct?"

"Yes," Avalon said. "I should learn that."

"I'm sure you will," Akenov said. "After all, you were born here. You are Kazakh. Here at school your brothers and sisters are studying English, and they will be happy to practice on you. And you can learn more of your native tongue."

Avalon held her hand over her cup when she saw Asel lift the teapot to refill. Asel smiled and motioned with her hand toward the door. Avalon nodded, stood and said, "Dr. Akenov, I hope you will have another interesting question for me. Kosh sow bolingdar."

Akenov stood and put his hand over his heart, and after a moment Avalon did the same. "Avalon, you amaze me. I will try to think of another interesting question for next time we talk."

Chapter 20

After closing the door to Akenov's office, Asel led Avalon down the hall to the left, opposite to the direction in which they arrived. At the end of the hall they descended a flight of dimly lit stairs, and at the bottom Asel pushed open a thick metal door. Light poured in from the playground. The door closed with a solid thunk and click behind them. The authoritative sound made Avalon turn. The green door had a simple handle and a keyhole. Asel led across the playground. The usual slides and swings, a roundabout and seesaw, a jungle gym, stood ready among scattered trees shedding rounds of shade. Avalon paused to study the meter tall numbers painted on the high walls:

6.62606957 e-34, 1.602176565 e-19, 3.14159265358979, 2.718281828459, 0, 1, 1, 2, 3, 5, 8, 11, 21, 34, 55, 89, 144... and at least 20 more such numbers.

Asel glanced back and saw that Avalon was no longer following her. Instead she was looking around at the walls, alternately frowning and nodding.

Asel asked, "What do you think of our number walls?"

Avalon smiled. "I saw numbers like these on the wall of Dr. Akenov's office."

"So, what do you make of these numbers?"

"Three point one four one five—easy enough, that's pi." Quickly in her mind the numbers separated themselves into groups and each group called up a name. "Um hmmm. And over there, the one beginning 2.718 is the value of the natural logarithm e. The others are fundamental constants of physics, like Planck's constant, and the charge on the electron. Two of them I don't know. And by the way, that 11 should be a 13." Avalon thought the mistake might be deliberate—to discover who would notice. Who *could* notice. Whoever put the numbers there, or ordered them to be put there, must have known the correct sequence.

Asel was smiling. She liked not only this girl's amazing intelligence, but her quiet self-confidence. "Well, you can tell Dr. Akenov when we see him again. Meanwhile, let's meet your roommates."

Avalon felt a pang of disappointment. This nice woman did not know the mistake was deliberate. Immediately after that pang of disappointment she felt a cloud of loneliness pass over her. Here she was alone not only in her way of seeing the world but in a strange country.

Asel's voice broke in. "They don't know you're coming, so it will be a big surprise for them. It's that door over there."

When Avalon started walking toward the apartment entrance, Asel looked back and saw Akenov watching through his window. She gave him the universal thumbs up signal. Avalon had passed another test. None of the Kazakh children more than glanced at the numbers when they arrived, and with few exceptions, none had even been curious about their significance. And only Avalon had pointed out the intentional error in the Fibonacci series.

The dorm complex occupied three floors, and the door to the first floor opened when Avalon tried it. She went in and immediately stopped when she heard a muffled but clear soprano voice singing the melody of one of her favorite Debussy pieces, each note exactly on pitch. Asel moved around her, saying, "That's one of your roommates. Her name is Mariam." She led Avalon toward the door down the hallway between walls painted with bright geometric designs in orange, purple and green. Asel knocked on the third door on the left. The singing stopped, and they heard footsteps pounding toward the door, which was flung open. "Asel! Asel!" Mariam threw herself into Asel's arms, hugging her tightly, then looked shyly past Asel's shoulder at Avalon.

Asel freed herself, turned and put a hand on each of Avalon's shoulders and said, "Mariam, this is Avalon. She will be staying with you for a while."

Mariam surprised Avalon by quickly coming to her and squeezing her with another enthusiastic hug. "Avalon!" Then she turned away and ran back to her bed, sitting cross legged in the middle and picking up the melody precisely where she had left off. Two other beds had been neatly made up and covered with purple and green quilts. A tall, broad wooden wardrobe painted bright orange stood in a corner.

"Mariam is always singing," Asel said, raising her voice to be heard. "But she will stop for a while if you ask her. Just say 'zuwas' firmly, which means quiet. Mariam's talent is music. She can sing the melody of a piece after listening to it just once. She also composes in her head, sings for anyone who will listen, but she can't read or write music. Dr. Akenov brought visiting musicians from the

State Orchestra here last week. They said her melodies are so inventive, yet so perfect in composition, that they recorded them and they'll use them in a new symphony."

She paused for a moment, smiling fondly at Mariam. "As I'm sure you noticed, Mariam is very friendly, very sociable. Everyone loves her, even though her singing can get on your nerves."

Avalon watched Mariam for a moment. The plump 10 year old had a full brown moon of a face and bare feet with thick calluses. Her long black hair hung freely to rest on her shoulders. Her shorts and blouse needed washing, and for that matter so did Mariam.

Asel walked to the wardrobe and opened the door. Under the shelves of neatly folded clothes Avalon's roll aboard bag stood next to her backpack. "Avalon, you can unpack now. The bed by the window is yours, and the dresser next to it. Your other roommate is Nurzhan, and she will be here after her tutoring session." Asel turned to go, then added, "If I see Nurzhan, I'll ask her to come by and say hello. Dinner is at six, so you might want to wash up and change. The bathroom is just down the hall, where you can shower if you wish. Nurzhan will show you where the soap and towels are kept."

Mariam stopped singing when Asel turned toward the door. She jumped up and ran over to hug her again. "Khosh! Khosh!"

"Khosh, Mariam." Just then the door opened, and a girl nearly as tall as Asel stopped with one foot across the threshold, surprised to see three people in her room.

"Oh good, here's Nurzhan," Asel said, stepping back to let her in. "Nurzhan, I'd like you to meet Avalon who will be visiting Good Shepherd for a while." Nurzhan glanced at Avalon, then back at Asel with a mixture of fear and puzzlement. Asel then explained something in rapid Kazakh, and Nurzhan nodded.

"Avalon, Nurzhan can help you settle in. By the way, she loves mathematics, just like you do, and knows a little English." With that, Asel stepped into the hall and closed the door, leaving Nurzhan and Avalon uncomfortably looking at each other. Nurzhan's face showed no emotion. Avalon felt as if she were facing a human camera, a web cam dressed as a girl robot. Finally Nurzhan said neither warmly or coldly, "Good bye."

Avalon smiled. "Maybe the word you want is hello?"

Nurzhan continued looking at her, possibly into her. Her eyes fixed, she said, "Zhaksa! Hello!" They both laughed. Mariam stopped singing and looked at them. She had no idea what the two girls were saying and felt she was being ignored. She got up, and yelled "Endi jehr!"

Nurzhan glanced at the clock on the wall, then at Avalon. "Mariam say, now eat."

Avalon had to laugh. "Well, I guess that settles it. And I am a bit peckish."
Nurzhan looked up at her. "Bit peckish? What is peckish."

"Peckish is a word my mother uses, probably from the time she spent in
England. Have you seen baby chickens, how they hunt around on the ground
and peck at whatever looks edible?" Avalon used her fingers on her palm to
illustrate how a chick pecks at food.

Nurzhan nodded. "I will remember. When I hungry like chick, I peckish.
Bit peckish."

Mariam had left the room and was waiting impatiently in the hall. "Peckish!" she shouted, causing Avalon and Nurzhan to burst out laughing. Mariam
had been listening to their conversation and picked up a new word.

When Nurzhan pushed through the glass door to the cafeteria, a dense
cloud of noise rolled over them. Avalon momentarily felt as if she were about
to drown. She quickly surveyed the scene, counting 20 tables with bench
seating, each with a dozen children ranging from 10 years to mid-teens, all
chattering. Kazakh music was playing in the background, and a long central
table held an assortment of cheeses, fruits and cakes, with hot meat dishes and
potatoes available at one end. It reminded her of the student cafeterias she had
occasionally visited with her mother at University of Oregon. Mariam picked
up a tray and headed for the potatoes and desserts. Avalon understood why she
was pudgy.

Nurzhan handed her a tray, taking one for herself, then fetched two Fanta
sodas from a small refrigerator and led the way to the hot food counter. She put
one of the sodas on Avalon's tray, and said "I order for you, OK?"

Avalon nodded, because she did not recognize any of the dishes. Nurzhan
spoke to the women on the other side of the counter in rapid Kazakh, and
watched carefully as two dishes were loaded with what looked like lamb mixed
with potatoes and rice. She passed one of the plates to Avalon, then checked to
see if Mariam had found a place to sit. She had, but the table was already filled,
so Nurzhan looked to the far end of the room, near the windows that opened
onto the now dark play area. One of the tables was empty except for an older
boy sitting by himself. She saw his hand move, then he checked something
that had fallen onto the table and wrote in a notebook. Nurzhan saw him too,
and said, "Good! Marat. You like him. He talk good English."

Another person to like, Avalon thought, was not what she really wanted.
She liked most people. What she wanted was someone like herself. She
surveyed the brown faces at the tables wistfully thinking she might discover
someone like herself. Why not even a brother or sister? She saw a few tall girls,
fewer tall boys, but no one looked like her except in color.

They carried their trays over and sat down, Nurzhan next to Marat with
Avalon across the table. Nurzhan said something to Marat in Kazakh, but he

ignored her. Avalon could now see what he was doing, and was puzzled. He would pause for a moment with his eyes closed, then write a single letter in the notebook and immediately toss a coin. When it came to rest on the table he would note how it landed and write another single letter, followed by a check mark or an X. Avalon nibbled at the unidentifiable meat dish while she watched him toss the coin and mark his notebook. After a few bites of the meat her mouth was coated unpleasantly. She concentrated on the rice and potatoes while sipping the one familiar thing, her soda.

She saw Marat draw a line under the marks and run his finger quickly down the line of check marks and X's. He wrote down 49 and 51, stared at the numbers for a moment, frowning. Then he surprised her by suddenly reaching across the table, grabbing her hand and shaking it. "Hello Avalon. I'm Marat Satpayev. Nurzhan tells me you're a visitor for a few days."

Avalon wiped her mouth with a napkin and smiled at him. "Well, I hope it's just a few days. What in the world were you doing with the coin?"

"Oh, that. I was testing one of my ideas about time—the Satpayev time theory."

Avalon immediately paid attention. The nature of time was one of the questions she often thought about, but without an answer. "What about time?" she asked.

"I was testing whether I could see into the future."

Avalon laughed. Marat, clearly miffed, began to gather up the coin and notebook, preparing to leave. She looked back down at her plate, using her fork to spear something that looked like a green bean. "I'm sorry, Marat. I didn't mean to make fun of you. If you get down to the reactions of fundamental particles, I think there can be a question about which direction time goes, but at the macro scale, because of entropy, our perception of time is unidirectional, which means you can't."

Now Marat was looking at her, not in amazement, like Asel. More like recognition. Then he stared, because he thought he saw one of her eyes flick as though to glance at Nurzhan while the other stayed focused on the notebook he was holding. But he wasn't sure, because both of her brown eyes quickly came back to him.

"But the line between past and future is not infinitesimally thin," he stated.

Holding his eyes, Avalon said, "Yes, so?"

"If you agree with that, it must be possible to plot the rate at which information becomes available to the brain. As the future becomes what we call the present, and the present becomes the past, data flux rises rapidly from near zero to a maximum then falls off as present becomes past."

Avalon understood, and nodded. She was beginning to enjoy the conversation.

"Think of the shape of the curve. As future becomes present, it does not instantaneously rise from zero to maximum flux. This means that I should be able to see a little into the future. The only question is how far."

"And the coin tossing?"

Marat shrugged. "It was an experiment. My brain and my thoughts exist in the present, but ten seconds from now they will be in the future. I can expect that my brain and thoughts will go there together, traveling forward in time. I told myself that every time I tossed the coin, I would send a thought back into the past, attempting to tell myself whether it came up heads or tails."

Avalon could not help grinning. "Did it work?"

Marat smiled ruefully, shaking his head and looking down at the coin on the table. "No. Out of 100 attempts, my guesses are what would be expected according to chance."

"There must be something wrong with your hypothesis."

Marat looked up. "What do you think it could be?"

"Simple. What we perceive as time is just the hundred or so milliseconds required for the physical and chemical reactions that give rise to consciousness in our brains. The future and the past are mental constructs with no existence in reality." She had no sooner said that than she realized it didn't even apply to her own life. Her reality was that she didn't know the future and she wished the present was more like the past.

After a moment of silence. Avalon noticed that Nurzhan had finished her meal and was looking back and forth between the two of them. Then Marat nodded.

"That makes sense. Who are you anyway? Everyone here is Kazakh and you look Kazakh but you're not. Why are you here? How long are you staying?"

"I am Kazakh, and I'm not. I live in America, but they won't let me go back. It's a long story. Basically, my mother and father, my American parents, adopted me from Kazakhstan when I was a baby, and a couple of months ago my mother was invited to a meeting in Astana. The organizers said I could come too, so I did. I had a great time. But when we were waiting for our plane home, we were stopped at the airport. They said my adoption was illegal. So I'm stuck here until someone straightens this out."

"But why are you here, at Good Shepherd?"

Suddenly a voice through a loudspeaker repeated urgently Жер сілкіну! Жер сілкіну! Avalon had no idea what the words meant, but after the second repetition, all the students slid off their benches and chairs and took refuge under the tables. When Avalon did not react, Marat reached up and grabbed her arm, pulling her down. "Earthquake drill," he explained. If you don't feel anything shaking, it's just a drill. We have them all the time."

Avalon asked, "You have earthquakes? Are they so common that you need drills?"

"Well, the drills are common. And sometimes an earthquake but never very big. At least not often."

Avalon remembered. "Yes. My mother says she felt an earthquake when they were here to adopt me." A bell sounded and all the students crawled out from under the tables and took their seats again to resume talking and eating. Avalon said, "We have fire drills, but we're never so fast."

"You have to be fast in an earthquake," Marat said. "They time us."

Avalon picked up where they had left off before the drill. "You asked me why I'm here at Good Shepherd. Dr. Akenov was the person who invited my mother to give a talk. I couldn't leave the country, so he offered to take care of me for a while."

"Wow! Dr. Akenov himself. That's impressive. Actually, my mother is his assistant. They work on problems about infertility."

"Your mother? You mean Asel?"

Marat nodded. "She and dad worked at the UN in New York before coming back to Almaty, so I was in school in Brooklyn from kindergarten to sixth grade."

"So, that's why you have such good English. Your mom too. I met her, by the way. She picked me up at the airport and we drove here together. She's nice."

Nurzhan was beginning to fidget, and interrupted, saying something to Marat, too fast for Avalon to follow. Marat listened, then turned back to Avalon. "Nurzhan reminded me that I promised to play chess after dinner. She's gotten to be very good, and I think she likes to beat me. Do you want to watch the slaughter?"

His question caught Avalon in the middle of a yawn. She shook her head. "I'd like to, but I haven't slept since the night before last. And I still need to unpack. I think I can find my way, so I'll see you in the morning." In fact, one part of her wanted to watch their game. Then she might know how smart they really were. The winning part of her didn't want to know—not yet while hope was still sweeter than knowledge. She quoted to herself:

HOPE is the thing with feathers
That perches in the soul,
And sings the tune without the words,
And never stops at all,
And sweetest in the gale is heard;
And sore must be the storm
That could abash the little bird

That kept so many warm.
I've heard it in the chillest land,
And on the strangest sea;
Yet, never, in extremity,
It asked a crumb of me.

Back in her shared room, Avalon was too tired to unpack. Instead she languidly dropped her clothes on the floor, then crawled under the covers of her bed and fell into a deep sleep even with the light on. When her two roommates returned an hour later, Mariam was singing as usual. Nurzhan saw Avalon asleep when she opened the door to the room, and whispered loudly, "Mariam! Zuwas!" They both tiptoed in, then prepared for bed as quietly as possible. Mariam couldn't help humming under her breath, but even an earthquake would not have awakened Avalon.

Chapter 21

When Evan Lipkovich called Michelle from the Eugene airport early in the morning, she took a minute to rethink her plan for his visit. She picked up the phone again and called Don Koskin. "You're on. He should be here in fifteen or twenty minutes."

When Don walked into her office he stood in the door a minute, and they looked at each other.

"Still yes?" he asked.

"Very much a yes," she said confidently, and hammered one of her imaginary nails with her fist. The next moment she felt embarrassed and uncertain as Don settled into a chair, crossed his long legs, bowed his chin to rest on the steeple of his fingers and looked up at her.

They agreed that Michelle would make the introductions. They also agreed that if Evan Lipkovich refused to work with the two of them, Michelle would not insist.

"You need them more than you need me," Don said.

"In fact, I've been thinking, and I don't really need them."

Don looked up with a question in his eyes.

"I'm sure that Akenov is the person behind all of this, and if somehow he's not, he has the power to end it. And I know how to make him do that."

Don again spoke only with his eyes, a question.

"Let's save that until later," Michelle said. "What's important now is that even though I don't need the CIA to help, I do need them not to become an obstacle."

Don uncoiled and leaned forward elbows on his knees. "Bottom line—keep them friendly."

Michelle's eyes went to her door and back to Don. "And you are a very friendly man."

"I hope I don't have to try very hard."

"I don't think you will. Actually, I kind of like Evan Lipkovich. He was probably a very good student. Maybe a good football player too."

As Don Koskin and Evan Lipkovich shook hands, Michelle noted that they were almost exactly the same height, although there seemed to be almost twice as much of Lipkovich compared to Don. Evan turned to Michelle, looking puzzled by Don's presence, but she answered his question before he asked. "You're going to have to accept one more person in the circle. I need to have my closest friend advise me, and he might be useful to you too. Please sit down."

Lipkovich hesitated a minute, but as soon as he had lowered himself into a chair, Michelle knew his answer was yes. She gave a quick summary of Don's qualifications as a friend and as a scientist who knew genetics and who had connections in Central Asia. When she finished, she opened both hands to Evan and said, "If your answer is yes, you can explain the rules."

After Evan had briefed Don and Don had signed the confidentiality agreement, Michelle deftly resumed control, starting with a charm initiative. "First Evan, I apologize for yesterday's phone conversation if I was a bit, what should I say . . . bossy maybe."

"Understood," he confirmed.

She settled into her chair. "So, here is everything that happened from the time we got to the conference until they forced me onto the plane to come home."

Evan took a pad from his briefcase, and as Michelle talked, he scribbled brief notes. When her story ended, he had filled one page and half of the next. Don, who was hearing that story for a second time, sat quietly, but was listening for anything new or something he might have missed.

"Questions?" Michelle asked.

Evan said, "I know your first priority is your daughter, and I know you told me on the phone not to expect a report until you had her back."

"Sorry about that," Michelle said. "I can organize some notes, but you'll understand if I don't write a lengthy formal report."

"Understood," Evan said. "Tell me very briefly if during the conference or afterward you learned anything or saw anything we should know about Dr. Akenov."

"Yes, I did learn something." She leaned forward, looked at Don then fixed her eyes on Evan. "I learned something this morning, or maybe I learned it while I was sleeping. Call it a revelation, or maybe just simple addition. I had the first part of the equation even before I went to the conference, thanks to Don."

She glanced up at Don who looked back without understanding, but said nothing. "The second part of the equation I brought home with me from the conference, without knowing it. This morning I added the two together."

Michelle paused for a moment to think how to make her revelation clear to Evan, because she needed non-technical language. She glanced at Don, who smiled and nodded. He understood her problem. Then she turned back to Lipkovich, who was sitting up straight and looking very large.

"Evan, no offense, but let's pretend you're a student in one of my classes. Anything you don't understand, just raise your hand and ask a question. I'll begin with your original idea that Akenov was developing a viral vector to deliver a pancreatic cancer oncogene, because that's part of the equation. This is certainly feasible. When I first began to work with TransTek I experimented with the SV40 virus, adding it to a human cell culture, and it was amazing to watch how it transformed orderly, well behaved fibroblasts into malignant cells that grew beyond control, piling up in the flasks as cancer. The problem with SV40 is that we couldn't find a way to insert other genes into it and turn it into a vector for delivering those genes to specific cells. So now I'm using a modified rhinovirus as a vector."

Evan raised his hand.

"Go ahead," Michelle said. She liked how Evan readily accepted her student-teacher Socratic dialogue.

"How much do you think Akenov knows about your work at TransTek?"

"Well, if he's responsible for the hacking, he certainly knows that I work there and the general nature of my work, but not the rhino virus work. I haven't yet created a report for the board of directors about the new vector. The results are kept as encrypted files in my new computer that doesn't have access to the internet, following your suggestion."

Evan knew that a determined industrial spy would have no trouble extracting and decoding computer files encrypted by an amateur, but kept that thought to himself. He wrote something on his pad as Michelle continued.

"The next thing you need to know is that evolution has tailored viruses to infect specific cells. For instance, cold and flu viruses target cells in the upper respiratory tract. The viruses adhere to the cell surface and wait for endocytosis to let them in."

Evan frowned. "Endocytosis?"

"That's the process by which living cells swallow bits of their outer membrane. Anything that is attached to the membrane gets in as well—a virus, for instance. Once inside they deliver their genes. A lot of my research at TransTek had the goal of engineering viruses to deliver genes that we wanted to insert into cells in order to cure genetic diseases. If I can do that, presumably Akenov might be able to design a viral vector to target healthy pancreatic cells. The virus would carry genes that activate pancreatic cancer."

Michelle paused for a moment to let that sink in. She knew what Evan was thinking. Pancreatic cancer had become famous by killing famous people— Steve Jobs, Luciano Pavarotti, William Safire and hundreds of others. Everyone knows that it is possible to make bombs that kill, but only one in a million human beings actually do it. Arman Akenov could be making an undetectable weapon to be used by sociopathic dictators to remove rivals.

She waited until Evan finished his note taking. "But I don't think Akenov is making a cancer gene. I suppose he might be, but that's not the primary goal of his research."

Michelle saw the surprise on Evan's face. "Then what?"

"Think about his horses," Michelle replied. "Akenov is famous for genetically modifying horses. He is immensely proud of his horses, the fact that they are winning races not just in Kazakhstan but in international competitions around the world, including the Kentucky Derby this year. He does this not by breeding, but by taking the short cut of cloning. He cultures cells from his fastest horses, then inserts their nuclei into ova and uses in vitro fertilization to produce genetically modified colts."

Michelle paused for a moment, looking at the two men. Don had a slight smile of anticipation. Evan had a wrinkle between his eyes that suggested he thought she was wandering. "Here's the deal," she continued. "Cloning is slow and pretty much hit and miss. With laborious cloning, he might get a few genetically modified colts, but a viral vector will make the process much more efficient. I'll bet Akenov tested the SV40 as a vector for delivering genes, basically the same approach we are using at TransTek. "

Evan said, "I still don't get it. All that work just to make some speedy horses?"

Michelle shook her head. "Put yourself in Akenov's position. He's a small man, not very imposing, but supremely intelligent. Everything he achieved comes from the fact that he is smarter than the people in power in his country. He finds it easy to manipulate them by playing on their sense of Kazakh pride with his horses. He has no social or emotional intelligence. He is basically a loner, bored by what to him is the dullness and stupidity of everyone around him. So, if you were Akenov, what's the most interesting thing you could do with your life? He knows that he can use modern biotechnology to improve Kazakh horses. What would be his next step?"

Michelle watched as the answer slowly dawned. "That's right. Humans. The Kazakh people. And that's why he arranged for Avalon to be kept there. She's Kazakh by birth, and her unusual intelligence may match or even exceed his own. I wish I'd never published this paper, because he certainly knows about the unique gene she carries."

Michelle picked up a reprint and handed it to Evan. "I didn't understand the significance of the gene at the time, ten years ago, but I think I do now. You came to my lecture where I described the first primate having normal human intelligence?"

Evan nodded.

"I called her Ma in my talk, and we think she had a mutation in the FOXP2 gene that changed something in her brain, probably the way that neurons are organized and interconnected through synaptic junctions. We don't yet understand how the gene changes the neurons and junctions, but it made human consciousness possible."

Michelle stood up and walked to the window, looking out at the green lawns and trees of the campus, now wet from a late afternoon drizzle. She looked over at Evan and Don. They saw the sadness in her eyes but also a steely determination.

"Before we left for Kazakhstan, Avalon told me she thought she was a mutant. To say it more scientifically, she has an important mutation. I think that Avalon is like Ma. Her unique gene represents the next step in human evolution, but it can only be maintained in the human genome if she has progeny. That's why she's being held in Almaty. For Akenov, Avalon is like breeding stock to be used to breed a super race of Kazakh people."

Michelle paused for a moment to let that sink in. Her listeners looked at her, considering alternative explanations, but none were as plausible. Evan asked, "Won't that take a while?"

Michelle nodded. "It would, but Akenov has a short cut based on his work with horses. I'll bet he's planning to get a DNA sample from Avalon and isolate her intelligence gene. Then he will try to insert that gene into the human race. Or more precisely, the Kazakh race."

Evan stopped taking notes and sat back in his chair, staring at Michelle in disbelief. "Is that actually possible?"

"Believe it or not, that's what I'm doing over at TransTek. Not intelligence genes, of course, but genes like Factor VIII that is missing in hemophiliacs."

Don broke in. "Evan, if you find that hard to believe, there's one more piece to the puzzle that I didn't know about until Michelle told me. It's crazy, but here it is. Guess what Akenov does besides breeding horses?"

"Well, his school for gifted children of course."

"Michelle, tell him what you told me."

"Evan, Akenov is the director of a fertility clinic just down the street from the school."

"So? I don't get it."

"He has an endless supply of young women wanting to get pregnant. One of the techniques commonly used is called ICSI, for intracytoplasmic sperm injection. If a husband has non-motile sperm, for instance, it is possible to harvest several ova from the wife and inject a sperm directly into the egg. The fertilized egg is then implanted in the wife's uterus and she undergoes a normal pregnancy."

Evan frowned. "I'm sorry to be so dense. Spell it out for me."

Michelle nodded, then stated what had become obvious to her. "Along with the sperm, Akenov could inject Avalon's intelligence gene."

Evan raised his eyebrows. "Can that really be done? Hard to believe."

"It is possible, but because of ethical concerns no one has attempted it yet. Or if they have, the results were not published for obvious reasons."

Evan looked over at Don, the back to Michelle. "You folks live in a different world. Anyway, I think I have enough information to digest for awhile. I'll go back, talk to Rhonda, see if we can come up with a plan. But there's something I need to do before I leave. Did you arrange for me to meet with. . . what's his name again. . . Bakit?"

"It's spelled B A K H Y T, pronounced bakeet with the accent on the second syllable. Yes, he's downstairs in the lab and knows that someone wants to see him, but he doesn't know why. You can use my office—I'll go tell him to come up, then while you're talking Don and I can have lunch."

"Sounds good. Before you go there's one more thing to tell you. Remember the napkin you brought back? You thought it might have Akenov's DNA on it?"

Michelle nodded, giving Evan her full attention.

"Well, there was some DNA and it all seemed to be from one person, so we assume it's his. We did a complete genome analysis, but there was nothing unusual except for one thing. Akenov definitely has the genghisid marker sequence. His paternal lineage goes back to Genghis Khan."

Chapter 22

The strong morning light spreading across her bed wakened Avalon. She turned away from the window and saw the figure of Nurzhan silently moving about the room.

"Hi Nurzhan. I'm awake."

Nurzhan replied in a whisper, "Shhh. Mariam sleep. Come." Avalon slipped out of bed, stretched, and picked up the towel and a small bar of soap that Nurzhan had laid out. She followed her down the hall to the lavatory shared by the students on their floor. This early they had it to themselves. Avalon hesitated before taking off her pajamas. She had never shared a shower or even momentary nakedness with anyone her age, but Nurzhan paid no attention to her, obviously used to shared facilities. Avalon turned on the shower head next to Nurzhan and relaxed as the steam built up around them. As she soaped herself she saw someone else undressing on the other side of the steam and when the singing began recognized Mariam. The pleasure of water and its warmth seemed to activate the music that played continuously in Mariam's head. Avalon smiled to herself at the inventiveness of the melodies that poured out. It reminded her of the mockingbirds in Eugene whose songs greeted sunrise every morning in the summer. She understood why people a century earlier had captured and caged the birds, but the passage of this through her consciousness made her wince. Mariam, the mockingbird, its history, her separation from Michelle formed a composition deep in her mind.

Avalon finished first and dried her body with the towel. She wrapped it around her mop of long, black hair and squeezed, but her hair still hung heavy and damp against her back as she made her way to the room. She missed her hair drier's little desert hurricane. From her suitcase she chose a fresh white blouse and jeans, the same blouse and jeans she wore at home in Oregon. For the first time in her life, she understood what homesick meant.

When Nurzhan and Mariam returned a few minutes later, they dressed in the school uniform for girls, a light blue blouse and a dark blue jacket and skirt. Avalon noticed the school emblem on the jacket, which looked like a stick with a floppy O on top. She pointed to it and asked "What is that?"

Nurzhan replied, "Urga. For animals. Not know English word. Baaaaa."

Mariam immediately imitated her. "Baaaaaa! Baaaaaa!"

"Goats?" Avalon guessed. "No, sheep! Oh I see. It's a kind of lasso stick."

"Zhakse! Sheep! Now eat."

Avalon followed them out into the courtyard to the dining hall where they joined a handful of other early risers for a simple breakfast of groats, toast and jam, orange juice and milky tea. Most were still sleepy, so it was much quieter than the noisy dinner the night before.

Avalon looked for Marat but could not find him.

"Nurzhan, where is Marat?"

Nurzhan shook her head. "Not here in morning."

"Where is he?"

"Live with mother."

That interested Avalon. She had had the impression that the school was enclosed by the wall she had seen, and that the students remained within the compound for security. Or was it to keep them from running away. The school didn't look anything like the children's home, the orphanage, they had visited in the city, but Avalon made a note to ask some of the students about parents—if they had any.

Marat's liberty added a dimension to student life that she liked. However, she did not yet grasp why he should be different, though she welcomed the fact. *Birds can fly* she said to herself.

After breakfast, Nurzhan and Mariam departed for the first classes of the day. With nothing to do, Avalon returned to her room. She noticed that Nurzhan had left another mathematical relationship for her on the whiteboard, not in numbers but in words: Even = prime + prime? She was puzzling it out when she heard a soft knock at the door. Before she could answer Asel Satpayev entered, wearing a white uniform and carrying a black leather handbag, obviously on the way to whatever work she did with Dr. Akenov.

"Avalon, good morning. You look more rested. I hope you had a good sleep."

"Yes, like a log."

"Log? I don't understand."

Avalon laughed. "That's just something we say in English when we have a good sleep. Logs, the trunk of a cut tree, lie as quietly and still as people would like to do when they're asleep."

"Ah! We say that we sleep like a baby."

"We say that too, so maybe I was a sleeping baby last night."

"I'm on my way to work, but wanted to stop by in case you needed anything."

"Yes, there is something I could use. I wonder if you could find an English-Kazakh dictionary? I know some Kazakh words from listening, but it would go faster if I could read them."

Asel was pleased. "Of course! We have dictionaries in the library."

"The school has a library?"

"Of course! You haven't seen it yet? I will show you. It's on my way."

Asel led Avalon across the playground to stairs next to the porter's lodge that went up to the second floor of classrooms. A few classes were already in session, and Avalon heard the voices of teachers. At the end of the hall Asel unlocked a door using a keypad. Avalon watched her fingers and stored the code, just four numbers long. Not that she ever intended to use the code, but remembering numbers was automatic for her. She had wondered before why the numbers on keypads were foolishly arranged in the same pattern on every pad so that she could know the numbers from the way fingers moved.

Avalon followed Asel and was surprised to see a small but modern library. One set of low shelves by the window was surrounded by comfortable upholstered chairs with reading lamps, and recent copies of journals like *Nature* and *Science* perched on a magazine rack. She walked over to the main area, three rows of tall shelves filled top to bottom with books. She scanned the titles. She could not read the Russian or Kazakh titles, but many books were scientific texts on genetics, molecular biology, and biotechnology. Within each category, the books stood in alphabetical order by authors. She found the neurobiology section and looked in the M's. She found it: *Genetics of Brain Development*, by Michelle Murphy.

By this time she was at the end of an aisle where a shelf marked Reference was given over to books with titles in Kazakh. Then she saw along the back wall of the room four study desks, each with a Macintosh computer that looked fairly new. When she touched a computer, she touched the whole world.

She called to Asel, who was waiting by the door, "Are the computers connected to the Internet?"

Asel walked over, nodding. "Of course."

"Can I use them?"

"Certainly. Our older students use them in their studies."

"Can I send an email?"

Asel paused. "That can probably be arranged, but it will have to be sent from Dr. Akenov's office. These don't have a link for email, because our students don't really have a need for it. Sometimes Dr. Akenov says, 'learn first, talk later.'"

Avalon immediately sat down at one of the desks, saw that the computer was on, pulled out the keyboard tray, and tapped the space bar to wake up the system. She was surprised when the first screen appeared in Kazakh. She looked up at Asel and said, "How do I get English? I want to type a note to mom. Can you send it for me?"

Avalon moved her chair aside to make room for Asel to reach the keyboard, where she quickly brought up a word processing program in English. Avalon immediately took over, and Asel watched as her fingers flew over the keys for a few seconds. She looked over Avalon's shoulder to read the note.

mmurphy@gmail.com

Hi mom! Did you get home OK? I'm fine. I've had a look around, and this seems to be a nice place. I already have a BFF. I've decided to learn Kazakh so I can talk to them. Otherwise what they say is like cryptic! Asel is Dr. Akenov's assistant, and she is helping me with the information I need. Her son Marat is my friend. He's smart! BTW I saw dad's horse here. It's a paperweight!

That's all for now. I miss you. Love, Avalon

Asel nodded, then took a thumb drive out of her briefcase and saved Avalon's note on it. After ejecting the drive, she stood and put her hand on Avalon's shoulder.

"It's very good of you to think of your mother. I'm sure she's worried."

"Me too. I hope this legal stuff won't take too long."

"With Dr. Akenov's help it will go fast, maybe just a few weeks."

A few weeks, Avalon thought. *What were the documents doing during 'a few weeks?* She didn't say anything.

Asel started for the door, then realized that Avalon was not following. "Time to go now."

"What about the dictionary?"

"Oh! I completely forgot." Asel went to the shelf marked Reference, picked out a thin volume and gave it to Avalon.

"This will do for a start. It's a beginner's guide to Kazakh and has a dictionary of common words and phrases in the back."

"Perfect. Thanks so much. By the way, can I use the Internet for a while now?"

"I'm sorry, but the library is only open from 4 to 6 for student use. I can't leave you here alone."

Avalon sighed, then rose and followed Asel out the door.

They parted ways in the courtyard, Asel heading toward Akenov's office.

"Thanks Asel!" Avalon called to her. Asel was fumbling for her keys by then, and just waved her hand.

Before heading back to her room, Avalon decided to take a few minutes to look again at all the numbers painted on the walls. She was puzzling over $7.297,352,5698 \times 10^{-3}$ when she heard voices behind her, and turned to see a gate opening near the stairs. Marat came though, talking to someone on the other side of the gate. She couldn't help herself. "Marat! Over here!"

Then Avalon felt her face get warm. She had read about blushing, but had never blushed in her life. Marat looked up, saw who was calling and immediately trotted over, grinning. He stopped a little farther away than most people stood for conversation, and for a few seconds neither could think of what to say. Then Marat gestured at the walls and numbers. "Have you figured them out yet?"

"Some. Have you?"

"It took a while, and the Internet helped. Usually I could just plug them in and the definition would pop up right away."

Avalon grinned. "That's too easy. I'd rather figure it out myself. Most of them I know, but I don't know that one yet, the one that begins 7.297."

"Are you kidding me? DIY? Why?"

Avalon shrugged. "They were just interesting. I'm still trying to decide whether numbers actually tell us something about the way the universe works. I mean, numbers are just mental constructs, aren't they? Like your idea of past and future. There are no numbers in nature."

Marat stared at her, something she was getting used to. Then he said, "Well, you should know that one. I think it's called the fine structure constant. It's a big problem in physics because no one knows what it means." Marat laughed. "When you figure it out, let me know. Anyway, I came for lunch today and have a class this afternoon. You hungry?"

Having lunch with Marat was a much better prospect than working on Nurzhan's puzzle back in her room.

The two got to the dining hall just as a woman opened the doors to the first arrivals. The main offering resembled pizza, so each took two slices with a simple lettuce salad on the side. Other students were sitting down in small groups, but Marat ignored them and led Avalon to a table near the windows at the far end of the hall, now bright with warm summer sunlight.

As they sat down, Avalon saw Marat look past her. He smiled and waved. "Salamatsys, ba? Tughan kuninmen!" She turned and saw two girls carrying trays to the next table. They giggled happily, and one of them stuck her tongue out at him, then sang out, "Rakmet, Marat."

Avalon watched them sit down and begin to eat. She was puzzled by how much they looked alike.

Marat said, "They're twins, in case you're wondering."

"So that's it. What did you say to them?"

"Just wished them a happy birthday. They turned 12 yesterday."

"That was nice. I can't get over how much they look alike."

Marat nodded. "Actually, there are quite a few twins here, even one set of triplets, but you wouldn't know it because only those two are identical."

"How do you know the others are twins?"

"Well, for one thing they usually sit together at dinner. They're all the same age, of course, and there's often a family resemblance. But mainly at the birthday parties you realize it's for more than one kid. There must be a couple of parties every month, and their parents always come with presents."

"When is your birthday?"

"Two weeks ago, in fact. I'm an old man, 14 years now. When's yours?

Avalon looked away, her eyes suddenly filled with tears. "Next Wednesday. I'll be 13."

Marat reached over and touched her hand. "Your parents won't be here, will they? Is that why you're sad?"

Avalon shook her head. She thought for a moment and decided she would, she could, she should, give him a piece of herself. "You know I was adopted by

Americans and they are the only parents I know about. Now only my mother is alive and she's in Oregon."

Marat sat back and watched her finish the cake. ""Now I'm sad too. Next Wednesday, right? We'll have a party."

Avalon laughed through her tears, then wiped them away. "You don't need to do that."

"I don't need to, but I want to. End of story. I'll tell my mom. She'll know what to do, because she keeps track of birthdays and arranges parties for the kids whose parents live too far away."

Exactly the opening Avalon was looking for. "Do some parents come to visit?"

"Yes, sometimes." He seemed to think about something then said, "Usually it's just one parent, someone's mother."

"What does your mom do for her work?" Avalon asked.

"She's mostly at the clinic across the street, running tests on the patients. I guess you would say she's a nurse."

"What kind of clinic is it?"

"It's for people who can't have babies. My mother doesn't talk much about it at home, but I have heard about aiviev and iksee. Not sure what those words mean."

Avalon grinned at him. "Those aren't words, they're acronyms."

Marat frowned. "What's an acronym?"

"That's where you put together the first letters of words as an abbreviation. IVF means in vitro fertilization, and ICSI is intracytoplasmic sperm injection."

Marat was impressed. "How do you know that?"

"Sometimes I visit mom where she works at the university. It's kind of boring just sitting in her office, so she lets me go to lectures in classes. I learned those acronyms in a physiology class when they got to human reproduction."

Marat looked at a light in the ceiling, then directly back at Avalon. "Tell me more. How does IVF work?"

"The woman is given special hormones that force her ovaries to make eggs. The doctor collects them right off the ovaries and mixes them with sperm from the father. This is all done in a glass dish, which is why they call it in vitro, Latin for 'in glass'. Anyway, the eggs are fertilized by the sperm, begin to divide, and several are put in the mother's uterus to make sure that at least one of the eggs begins to grow into a baby."

"Several? Does that mean that more than one might start to grow?"

"Oh, for sure. One woman was in the news because she had eight babies. They called her the Octomom."

Marat thought for a moment. "Maybe that explains it. Why there are so many twins here."

"Really? Oh, I get it. The parents are treated at the clinic and lots of them have twins, then their kids come here when they are old enough. But why don't they just go home?"

"These are rich people who use the clinic, because it costs a lot. Good Shepherd is also very expensive, but it's right across the street from the clinic and it's safe, not to mention one of the best schools in the country."

Avalon nodded. "So why do you come here? Are you one of the students?"

"Not exactly. My mother lives nearby, and she doesn't need to pay the tuition since she works at the clinic. I just sit in on classes that interest me and use the library, which has Internet connections."

"Your mom showed me this morning. She said I could get on the Internet between 4 and 6."

"Well, it's pretty crowded then and each student only gets ten minutes, but Mom gave me the key pad code so I can use the library after hours."

There was a small beeping sound, and Marat paused for a moment, looking down at his mobile phone and then punched numbers. Avalon studied his face while he was doing this. Marat had straight black hair that came to a sharp point on his forehead, a characteristic she remembered was called a widow's peak. His lips were full, and his nose was not small and flattened, as she had noticed on many Kazakh people, but was thin with a slight hook. A word came to mind: aquiline. It made his face fairly easy to recognize. 'Aquiline and Avalon' she mused to herself, though she did this many times a day with word play.

"Just a message from mom." Marat glanced at his watch. "Gotta go to my class at noon, then home. Will I see you tonight? I'd like to catch up on news from the States. All we get here is stuff on the Internet, and it's been kind of filtered."

"Sure. What's your class? Maybe I'll come along."

"I don't think you'll find it very interesting. All the students here are required to take a course in politics and government, but it's mostly Kazakh-style propaganda. My mother says that I should make sure the teacher sees me there, because she works with Dr. Akenov and it would look odd if her kid didn't show up for a class that Dr. Akenov thinks is important."

"Well, I don't have anything better to do. At least I'll see what the classes here are like."

After placing their trays on a moving belt that carried them off for washing, Marat led the way up the stairs to a classroom that turned out to be one door removed from the library. In the hallway, Avalon noticed one of the men who wore the black leather jackets standing by a doorway. "Do you know that man in the leather coat?" she asked Marat.

Marat looked briefly as they walked away. "Not him, but they're always around. Lots of places in our country have guards, usually with guns. These men work for Dr. Akenov."

"He is watching me," Avalon said. She decided to say nothing more.

"They're guards. They watch everything," Marat said matter of factly.

Chapter 23

The class turned out to be as boring as Marat had predicted. Marat sat in the first row of seats with Avalon beside him, but as far as she could tell the elderly teacher never noticed her, even though he went through a roll call for attendance at the beginning of class. In fact, he hardly looked up from his tattered notes as he began his lecture in a droning voice entirely in the Kazakh language. Anyone looking at Avalon would think she was paying rapt attention, but in fact her attention was not fixed on what the teacher was saying. Instead she was listening to the language, and the sound of it ran through her head like something she could almost feel and see, not just hear.

The teacher's monotonous reading of his notes slowly faded away as Avalon descended into her inner world, thinking about her conversation with Marat the night before, and his experimental attempt to predict the outcome of coin tossing. Marat had made what she knew was a false assumption—that past and future actually existed. Avalon was confident that past and future were boxes into which we sorted things we knew about and those we only guessed at. She thought of the present as simply the focus of the ten billion neurons of the brain—what we pay attention to.

Even as she processed the sounds and rhythms of the teacher's voice, she pondered the question of how her brain, or any brain for that matter, worked. And that was the key to a more important question—who am I? Which led to a second question—why am I? Had the world simply reassembled itself 13 years ago into a girl who would be called Avalon? The world, or nature, of course, followed certain rules, but how had those rules led to such different things as the great white wall of the Tien Shan mountains that rose from the earth south of Almaty, and that little brown baby born somewhere in Kazakhstan?

And how could those rules determine that certain substances would become Avalon? And why was Avalon thinking about Avalon? Had she directed her brain to do this, or had her brain been destined to do it? If she could direct her brain to do this—wasn't that mind over matter? Her mind changing the physical components of her brain?

She had read about the brain, seen pictures of neurons and their dendrites and the synapses. She knew the brain was something like a computer that worked not on chips with binary logic, but instead neurons using energy to produce minute ionic currents through cell membranes. She could see it all working together, but she could not imagine herself as nothing but the totality of "things." It occurred to her that if she knew the answers to who and why she was, she might be able to understand what was happening to her now and how she could go home. She imagined a future that held infinite possibilities. She knew that her brain used sensory neurons to detect local possibilities around her, then filtered these through a funnel called the present, a tenth of a second of consciousness in which she could manipulate possibilities so that they became probabilities. A tenth of a second later everything became frozen into a finite number of outcomes called the past.

Her thoughts then turned to the question of why her brain seemed so different. She had often played with the idea that she was a mutant, and once she even got her mother to agree. Maybe her mother knew something and hadn't told her yet. Avalon recalled what Michelle had said in her lecture, that intelligence in mammals could be predicted by a simple mathematical function that added up neurons and synapses. Suddenly Avalon's thought process focused on that. It wasn't just the numbers of neurons and synapses. Instead the equation multiplied the logarithms of the two numbers! Differences in log functions seem small, but they can hide huge difference in the actual numbers. All it would take to explain why she was so different from others is a mutation that increased one of those two numbers—more neurons or more synapses.

Avalon's reverie was interrupted when Marat nudged her with his elbow. Her attention returned to the classroom, and she realized that the tone and pace of the teacher's voice had changed. He was looking at her and talking to her. In a low voice, Marat said, "He wants to know why you are staring at him."

Avalon said to the teacher in Kazakh, "I listen." The grammar was not quite right, but her ability to speak Kazakh surprised the teacher. Marat smiled and spoke in English. "Sir, this is Avalon, a visitor from America. She doesn't understand our language yet, but she is learning quickly."

"Why is she here?" the teacher asked sternly, also in English.

"Sir, she was invited by Director Akenov as his personal guest."

At the mention of Akenov's name, the teacher blinked and glanced back and forth between Marat and Avalon, then folded his notes and said something in Kazakh. This was followed by noises of rustling papers and scraping chairs. Avalon looked around to see students getting up, glancing over at her and beginning to talk with each other. Marat also got up. "Okay, let's go. He dismissed class early."

"Because of me?"

"Well, he noticed you were looking at him while most of the other students were either napping with their eyes closed or busy pretending to take notes. He is used to that, but the fact that you seemed to be paying attention bothered him."

"I wasn't looking at him. I mean, I was, but not really. I was listening but also thinking about something."

"What was that?"

Avalon thought for a second, then said, "Why some people are smart and others are not."

Marat burst out laughing. "Mr. Alizhanov was explaining why we all must obey our glorious leaders, and you were wondering why you are so smart."

In the hall outside a chattering, laughing parade of students hurried to the next class. Marat and Avalon joined a smaller group walking downstairs into the central quad where they shaded their eyes against the bright afternoon sun. Marat's phone chimed and he stopped under a tree to answer. He glanced up at Avalon as he listened, then said "Okay." Avalon raised her eyebrows, curious who was calling.

Marat shrugged. "That was mom. The Director wants to see you again."

"Why?"

Marat pointed across the quad with his thumb, "Haven't a clue, but she's waiting over there to take you to his office." Avalon looked and saw Asel in her white uniform holding a door open, the same door she had passed through yesterday after her first visit with Akenov. As they walked over, Marat asked, "What's he like, anyway?"

Avalon was surprised. "Don't you know him?"

"Never met him. Students only see the Director in the fall when he welcomes us at the beginning of the school year. It's an assembly in the auditorium, so we never see him up close."

"But your mother works for him!"

"So did my father, but he doesn't live with us."

Avalon, who seldom considered sensitivity before yielding to curiosity, asked,

"Why doesn't your father live with you?"

"I don't know, really. He has never lived with us."

They reached Asel, who thanked Marat and smiled at Avalon, then motioned her through the door.

Impulsively, Marat asked, "Mom, can we have Avalon over for dinner some night?"

Asel was surprised by Marat's question and hesitated. "We'll see," she finally said. Then she let the door close and led Avalon upstairs to Akenov's office.

As they walked down the hall, Avalon asked, "Did you send the email to my mom?"

"Well, I gave it to Dr. Akenov. You can ask him now."

Asel knocked on the office door, opened it to let Avalon in but did not enter herself. Avalon stopped just inside the door and saw that the little man was standing at the window overlooking the quad. She realized that he had been watching them.

"Come in, Avalon, and make yourself comfortable. Asel, please wait for us next door. We'll be there in few minutes." Asel nodded, then carefully closed the door. Akenov walked over to the couch and sat down. Avalon, who had been standing, sat in the chair she had used before.

"Well Avalon, what do you think of our little school?"

"Your little school has a very unusual little student body."

"What do you think of your roommates?"

Avalon could not help grinning. "If I gave them English names, I'd call them Miss Numbers and Miss Music. And I like both of them."

"I thought you might, and I chose them for you. I want to be sure you are comfortable here and have everything you need. Is our food giving you any problems?"

"Not at all. I feel very good," Avalon said, and indeed she did. Except she would rather be in Oregon. She didn't say that.

"Perfect," Akenov said. "And that is how we want to keep you. I hear that you and Marat are enjoying each other's company."

Avalon took notice—Akenov knew about Marat, but Marat had never visited Akenov. "Yes, he's nice."

"What do you talk about"

"Oh, lots of things."

"Can you be more specific?"

"Well, last night at dinner we talked about whether the future was real, whether we could see things happening in the future."

Akenov smiled. "That must be one of those interesting questions you like."

"Not really. What we call the future is just imaginary. That's what I told Marat."

"And what do you imagine is your future?"

Avalon was not sure if he was playing with her words or if he was asking for her opinion. "I will be a teenager for a few years," Avalon said. She looked carefully at his face for any sign of anger. She saw nothing. She had imagined many things in her future, but she knew the future was imaginary. A familiar slogan came to her mind. It fit both her sense of the future and the truth of her ambitions. "I want to be all that I can be," she said.

Akenov nodded. "It's good answer. And what you can be interests me very much, because what you are already is very different from our students. Do you know that?"

Avalon did know that but she only shrugged. "By the way, there is something I want to ask you," she said and leaned toward him. "I gave Asel an email note for my mom, and she said she gave it to you to send. Did it go out okay?"

"Of course. I sent it this morning, but she hasn't replied yet. I'll check again while you are here." Akenov got up and went to his desk where he tapped the space bar to wake up his computer. He began to scroll down through messages. Avalon watched. The light from the computer screen illuminated his face. A word suddenly came to her mind. *Aquiline. He has an aquiline nose.* Then Akenov straightened and walked back, shaking his head.

"Nothing yet. Of course, early morning in Oregon, so your mother probably still asleep."

When Akenov sat down again, Avalon studied his face. Yes. His black hair came to a widow's peak. And full lips, just like Marat.

Akenov noticed her scrutiny. "Something wrong?"

Avalon thought quickly. "You have something white on your chin, maybe shaving cream?" She pointed to the left side of her own chin.

"That can't be. I use electric razor." But he did wipe his hand over his chin.

"OK, you got it," Avalon said.

"Well, now that I have cleaned up, I have boring question for you. What vaccinations have you had?"

Avalon shrugged. "Vaccinations? What do you mean?"

"Shots to prevent polio, measles, chicken pox and other infectious diseases."

Avalon shrugged again so Akenov continued. "Your records say diphtheria. smallpox, measles and polio shots as infant in Children's home, of course. But that was years ago. I am specialist in vaccines and viruses. Here we use only best, including new vaccines I make."

"Do you follow recipes for your vaccines? Are you a good cook?" Avalon asked. The cooking-vaccine creation had just popped into her mind and out of her mouth, and she watched his face for irritation. She saw only a brief narrowing of the eyes in puzzlement. It was very brief, but she realized this man was like Emilia Bedelia, the children's book character who took every sentence at face value.

After a moment, Akenov continued. "Yes, I'm good cook, as you say. Now, I'm sure your mother would have given you all usual vaccinations for American children. But Kazakhstan is not United States, and we still have health problems left over from our past. You are not in great danger, especially here at

Good Shepherd, but you do not have immunities and resistances that our people have."

"You mean I'm an Indian and Kazakhs have small pox?"

Akenov stared at her. "You are not Indian. You are Kazakh and small pox doesn't exist anymore. Isn't it wonderful—after ten thousand years, we have killed this disease."

Akenov's inability to grasp her metaphor surprised Avalon but confirmed her comparison to Emilia Bedelia. She explained, "I didn't really mean Indians and small pox. That happened between American natives and Europeans. I just meant I'm a stranger and Kazakhstan might have new microbes."

Akenov put on his strange smile. "Yes. Now I understand. And I'm glad that you understand because we have pertussis here in Almaty."

"Pertussis? What's that?"

"Pertussis is whooping cough. Cause is highly contagious bacteria called Bordatella pertussis. Even if you received vaccination as child in Oregon, it's good idea to have booster vaccination when there is chance of exposure. Children here have had their vaccinations, I hope you will have one too."

Avalon thought for a moment. She remembered the one unpleasant week she had spent getting over the flu and decided that a booster shot against pertussis was a reasonable precaution. "OK."

"Excellent! Asel is nurse, you know, so she can give vaccination right now, then you can get back to your new friends."

Arman stood and led Avalon past his desk and work area to a small door. He opened it, and Avalon was surprised to see what looked like an extensive laboratory. From her occasional visits to her mother's lab back home, she recognized a sterile hood for tissue culture work, an incubator, a large refrigerated centrifuge and an Illumina DNA sequencing device. There was also a refrigerator and freezer, a confocal microscope and an atomic force microscope. Asel was seated at the confocal microscope which had a large image of a cell displayed on its computer screen. Cellular structures were mostly in shades of gray, but two bright green spots could be seen in the nucleus. Akenov strode first to the refrigerator and took out a small glass bottle, then returned to peer over Asel's shoulder at the screen.

"So. It worked."

Asel nodded, then added in Kazakh, "The new viral vector is just as efficient as the MLV strain, and seems to be significantly faster. We exposed the culture only yesterday afternoon, and you can already see both genes incorporated in the chromatin."

Their conversation continued in Kazakh. "Is that equine culture?" Akenov asked.

"No. Human foreskin cells, mostly fibroblasts."

"How old was child?"

"Just an infant, about three weeks."

Akenov switched to English, not realizing that Avalon had understood an important part of their conversation. "Avalon, you see here living cell."

"It's beautiful."

Akenov smiled at her and nodded. Avalon noticed that his smile this time was genuine, the whole face. Akenov was indeed happy to know she appreciated the subtle beauty of even the simplest form of life. He turned to Asel and set the small bottle down on the desk, along with a small syringe in a plastic container. "Pertussis vaccine. Give Avalon shot, then bring vial and syringe to me for disposal."

He turned to leave, but Avalon had a question. She pointed to the computer screen and asked, "What are the little green dots in the nucleus?"

Akenov was surprised that the 12 year old knew what she was seeing in the image on the screen. Her mother's influence, of course. "DNA bits we added to cell."

"What kind of cell is it?"

Asel noted a very small narrowing of his eyes and a short hesitation before he said "Horse."

A few minutes later, Avalon had a small round bandage on her arm. Asel returned the empty vaccine bottle and syringe to Akenov, then led Avalon downstairs and out to the play area. "Leave the bandage on for awhile, but you can take it off tonight. Your arm will be a little sore."

Avalon nodded, and began to walk across the quad toward her room. She was thoughtful. Why had Dr. Akenov lied? She was sure, perfectly sure, that Asel had said the cell was human.

Chapter 24

Shortly after founding the Institute for Genomic Research and its associated fertility clinic, Akenov had hired Sergey Nurlanovich Smagulov, one of Kazakhstan's best translators. When offering him the job, he said "There is one condition. You must have Kazakh name. Here you will be Nurlan, like your father." The thin, passive young man wore thick glasses. His hair was already thinning on top, and he owned just one baggy suit. He had not protested. His pay as a science translator and personal secretary was too good. His will was too weak. And as soon as he had been offered the job by Akenov, he knew this strange man had powerful connections. Besides, keeping a government job was easier for Kazakhs than Russians. His mother agreed that the money was too good to turn down. "As long as I don't hear that name," she sighed. "Does he know how lucky he is to have you—the best translator in Kazakhstan?"

"I'm not the best, Mother," Sergey said softly, with a tired but appreciative smile.

His mother kissed him on the forehead. "I never thought the day would come when my son could work for a Kazakh boss and make that much money!" Even though Akenov used him and abused him like a slave, he paid him like a movie star. He and his mother had moved to a modernized apartment in the village to be near Akenov's institute, an apartment with a high class "Euro remont" makeover that Akenov had secured for them. At the Institute, Sergey occupied a small windowless room with a fast computer that was never turned off, internet access, and several scientific dictionaries. His name was on the door in gold letters—Nurlan Smagulov. When Dr. Akenov wanted him, the computer played two bars from a familiar Kazakh melody and an amber light at the corner of the screen was programmed to blink. Sergey, now Nurlan, was not bothered by the thought that he responded to the light and the melody like one of Pavlov's dogs. After all, the dog received a tasty reward. Nurlan had found his guiding star long ago in an English history book. Prime Minister Lord Robert Cecil's son asked him, "Father, what really matters?" The Prime Minister replied, "Few things matter very much, and most things don't matter at all." And that was how Sergey lived.

A week after Avalon's vaccination, the light blinked, the melody played and within a minute Sergey, transformed into Nurlan the translator, knocked on the door to Akenov's office and announced himself, "Nurlan".

The routine had never changed. Akenov opened the door, Nurlan stepped meekly across the threshold. Akenov told him, "Sit down. I unload." When Nurlan had taken his position at a small desk at the other end of the office and turned on the computer that only he used, Akenov lay down on his large upholstered sofa and closed his eyes.

Usually he began to dictate notes in English, and Nurlan would type into the computer, occasionally making a note on his pad about something to verify. At the end of a session he followed the standing instructions and put the used notebook pages in the shredder. When he was done, he would take nothing from the room except his memory and the blank pages of his pad. His memory was very good.

Today Dr. Akenov had a warning for Nurlan delivered in Kazakh. The warning was familiar, repeated from time to time almost in exactly the same words. "You will remember that my work is very important. You may think you understand it, but you do not. That is why you should never talk to anyone but me. Correct?" Nurlan turned and saw that Akenov had not moved on the sofa or even opened his eyes. Nurlan said very agreeably into the silence, "I am not a scientist. I don't understand your work." That was true. He did not know anything about genetics or neurobiology, but every educated person in Kazakhstan knew about Akenov's contributions to agriculture and how he had made Kazakh horses the marvels of the equestrian world.

Then came what Nurlan knew was a serious threat. "My work is national treasure, but secret treasure. To give this to anyone else would be treason. You understand."

The last sentence was also a command. "I understand," Nurlan confirmed. And he understood that even his life might depend on this understanding. He had grown accustomed to bullies in school, and he had learned long ago how to be invisible in a crowd.

Dr. Akenov cleared his throat. A few seconds of silence passed before he said, "Begin."

Nurlan typed the date with a few key clicks. As soon as he stopped, the dictation began in a monotone voice.

"Biopsy taken with vaccination needle then retrieved viable skin cells from 13 year old Kazakh girl, subject 2000/NA-2B. Girl's fibroblasts grow well. They are in third sub-culture, four flasks ready for next step."

Long ago Nurlan had decided to type verbatim, and not to correct Akenov's omission of a, an, and the. He thought Dr. Akenov was not so smart about language. He was familiar with the coding, however, and asked, "2B or 2A?"

Akenov snapped, "I say 2B. Write 2B." He continued. "If cells respond to treatment and revert to iPS cells, next steps are . . ." He paused and said to Nurlan, "Make list now."

"List," Nurlan said.

"One. Initiate differentiation into ova and begin implantation trials. Procedure follows cloning protocol developed by group of Michinori Saito.

"Two. Saito paper omitted one detail essential, maybe so others could not reproduce their work quickly. I follow my procedure with horse clones. Same step I am sure they did with mice. Missing step is adding embryonic somatic cells from aborted female fetuses."

Akenov paused. He looked intently at Nurlan, who sat with fingers poised above the keyboard. Then Akenov continued. "Murphy paper in Brain Research reported unusual marker sequence in one child with possible autism. Child must be subject girl. Similar but not identical marker in two children here. One is classical Williams syndrome child who displays remarkable musical talent. Other is autistic mathematical savant. Last year new paper reported novel transcription factor in child with extreme mathematical ability. When compared to marker sequence in Murphy child, same except for tyrosine to tryptophan single nucleotide polymorphism. Interview with subject girl suggests unique mutation. Girl is pure Kazakh. Such Kazakhs will be new species. Caution: must confirm if her character from gene or education."

Chapter 25

Two days later Sergey Smagulov made a very difficult choice when his mother returned from the city's bustling Green Bazaar with her potatoes, tomatoes, cucumbers, Korean shredded carrot salad, mutton, and sour cream. "The pharmacy called," she said. "I brought home your prescription." She put a small jar of pills on the kitchen table. "Now what is wrong?"

"Just a little ear infection," he said. He took the pills and went to the bathroom. He opened the jar, shook the 20 pills into the palm of his hand and picked out one that was slightly larger than the others. It unscrewed in two halves. Inside one half he read the number 20. Inside the other half the word merlot.

Two hours later at 8 p.m., he stepped into the aluminum and glass street kiosk that went by the code name "merlot." The kiosk was crowded but he shouldered his way to the cashier and bought the bottle of wine, holding it in one hand so that it was visible. As he stepped onto the sidewalk a short young man with a trim black beard asked directions from him in the usual Russian way but with a detectable accent: "Excuse me, do you know this address?" The man held up in front of Sergey a crumpled slip of paper. Sergey studied it for a minute, looked the man in the eye. They had met several times before. Sergey was supposed to know him only as Maxim, but from the American embassy's publicity and web pages Sergey had easily identified his official position and name as Carl Venters, Export Specialist in the Economics Section. Although the embassy had moved from Almaty to the new capital in Astana, the Americans maintained a large consulate in Almaty because it was still by far the country's largest city and its economic capital.

Sergey said, "Let me see that again."

The man held it up again and said almost aggressively, "You must know this."

Sergey looked down the street and said, "Yes. Yes, now I know it. Take the number 28 down Abay to Rosabekeyevo. Three blocks south."

"I will return the favor one day," the man said as he lowered his hand with the address and crumpled the paper into his pocket. He hurried away toward the tram stop.

Yes meant Sergey had read the message which was not an address. He now had a mission that could end with a bullet to his head if he made a mistake and was discovered. This possibility, of course, meant he had become an important person. His new importance was not a fact he would want anyone to know, but it was a fact that he enjoyed. Sergey would like to be able to talk to that American he had met at the Embassy's annual 4th of July picnic, the big young man named Lipkovich. But he had seen Lipkovich only that one afternoon a year ago when Lipkovich's car brought him through the gates of a private residence. Sergey had signed a simple contract, promised not to use his name or to try to call him. He had refused a cash payment, instead asking for a bank account in England. He had dreamed of living in the land of Shakespeare and Shaw one day. Over the year he had delivered to Maxim occasional summaries of Akenov's work jotted down from memory.

But now, for the first time he would use his memory to take information into the Institute, or more precisely to the school, and this was potentially very dangerous. He had not met the American girl, but her picture accompanied her file in his computer. She seemed to be exceptionally smart, which for some reason had attracted Akenov's attention.

The next morning Sergey followed his usual routine. As he walked along the quiet streets shaded by the giant poplars, he thought of himself as an actor walking between the curtains that formed the wings of a stage, the grounds of the Institute. As the guard in the little kiosk at the gate noted his name and time on the log, Sergey momentarily pondered a name for the drama, but could not decide if it began "The Comedy of . . ." or "The Tragedy of . . ." Nor could he decide if his role was the fool, the villain, or . . . no, he would not be the hero. Never mind, he was a well paid actor whatever his role, and he should concentrate on being sure the other actors did not trip him up.

He made a special effort to greet everyone he passed. He settled at his desk as he thought he always did, but he was now oddly aware of studying himself and his actions as if he were not just an actor, but an audience as well. He did not like the script, but he played the role well. He opened several journals on his desk, booted his computer and began translating an article. He worked for 30 min, decided not to stop at such a precise half hour and worked another 11 min. He walked to the main science library shared with the Institute where he would take what he hoped was a small risk. He had stopped at the librarian's desk and was about to ask her if the American girl came in, but then saw the girl walk into the reference section and had to suppress a smile. He did not believe in a god or gods, but if they existed they were on his side this morning.

He walked slowly to the reference section, took a small notebook from his pocket and began to look at specialized English dictionaries. He picked out a medical dictionary, and sat at the table opposite the American girl. She looked up briefly and said in clear, deliberate Kazakh, "*Salam aleicham.*"

Sergey returned the greeting, opened the book and scribbled a few notes. Without looking up he said very quietly in English, "Don't look but listen carefully." He saw a slight adjustment in her hands and posture, but she did not look at him. "Your mother sends this message: Trust no one. Especially not your guardian. Help coming soon. Turn two pages, if you understand."

Avalon turned two pages. Sergey pushed back his chair and closed his book, about to get up. Avalon whispered without looking at Sergey, "Who are you? Why should I trust you?"

Lipkovich had asked Michelle to give him two details that Sergey could use if Avalon did not trust him. Sergey said them as they had been on the contact's message. "Your father's picture is on the mantel; your friend is Uncle Don."

Avalon said, "When . . ." but Sergey whispered, "No. I don't know more. I have my own danger." He put the book back with as much speed as he could without attracting attention and pretended to look at a few titles as he drifted out.

Avalon wanted to know what help was coming, when, and how. She had already sensed danger, and now this man said she was in danger. She knew that the security and order and peacefulness of the school and the institute were part of the danger. These qualities were not gifts. They were imposed by a dangerous power.

As she walked across the playground to her room, her mind worked fast, connecting random observations and what she had just heard. Like the pieces of a jigsaw puzzle, they began to form a mental picture, an understanding.

It was Friday, the end of Avalon's first week at Good Shepherd. She sat with Mariam and Nurzhan in the cafeteria, picking at her food, half listening to Nurzhan explaining how every even number can be expressed as the sum of two prime numbers, such as $8 = 5 + 3$, $14 = 7 + 7$ or $11 + 3$, and so on. Nurzhan had discovered this herself last year, and was a little disappointed when Avalon told her that someone named Goldbach had published this conjecture over 200 years ago. But Nurzhan brightened when Avalon added that no one had proved it yet. While Nurzhan prattled on about how she could try to prove it, Avalon was waiting impatiently for Marat to arrive for his afternoon classes. In the puzzle she had put together something that frightened her, something she needed to talk about before thinking much further.

Avalon had just started nibbling at dessert, some kind of sweet cookie, when Marat came through the gate and started walking toward the cafeteria. Leaving Nurzhan in mid-sentence, she jumped up and ran outside to meet him. Marat grinned and waved, but stopped when he saw her face. She was staring at him as if were wearing a mask or had something strange stamped on his forehead. "What's wrong?"

"We've got to talk. Is there somewhere private?"

Marat thought for a moment. "Maybe the science library? No one will be using it this time of day."

In fact, when she and Sergey had left, no one but the librarian had been there, and the librarian would be at lunch. "That's perfect. Let's go."

They sat in the back at a table near the emergency exit that opened only from the inside. Avalon drew several sheets of paper from her bag and put them face down on the table like a secret hand of cards. "I think I know why there are so many twins here," she said.

Marat's eyebrows went up but he said nothing.

"Have you noticed the way their hair grows?"

Marat thought Avalon's question very strange. "No. What do you mean?"

She turned over her papers with one picture on top, a quick but recognizable sketch of one of the twin boys they both knew. She traced the hairline over the forehead with her finger, making a wide and shallow V with curving wings. "That's called a widow's peak."

Marat grunted and said, "Okay, I have one too."

Avalon nodded. She spread out eight sheets of paper, each a sketch of a student in the school, each with a widow's peak hairline. "Yes, you have one too. It's one of the dominant traits that gets passed on from parents to children. Does your father have hair like that?"

"I don't think so, but it's not something I paid attention to. Anyway he's bald now."

"Okay, let's go to ears. I noticed something else about the twins here. They all have ear lobes that are attached, rather than hanging free."

Marat frowned. "Wow. It's weird you would notice something like that." He looked at Avalon's ears. "Yours are unattached, right? What are mine? I've never noticed."

"Yours are attached, like all the twins. It's a dominant trait."

"I still don't get it. What are you driving at?"

"There is one person here at Good Shepherd who has a widow's peak and attached ear lobes."

"Who's that?"

"The Director. Akenov."

Marat was smart. Very smart. In seconds the significance of Avalon's words sank in. "That's crazy! He couldn't be the father of all those twins!"

"Don't forget. He runs a fertility clinic. It would be easy to use himself as a sperm donor."

Marat touched his ear lobe. "But would that mean. . .?"

Avalon met his eyes. "Yes, he might be your biological father. A widow's peak is unusual among Kazakhs, and for you to accidentally share two such dominant traits with someone else would be very rare."

Marat stared at her. "That can't be," he said flatly. "I have a mother and father, and they were married a year before I was born."

"Was she working for the Director then?"

"No, but my father was. Mom was a nurse at the hospital."

Avalon was silent for a moment, looking down at her hands. Then she said, "There's one way to find out. You need to ask her."

"Ask her what?"

"If she was treated at Akenov's fertility clinic."

Marat was silent, shocked at what that implied.

"I wouldn't know how to ask her that."

Avalon thought for a moment. "Well, you can do it in a roundabout way. Ask her when and how she first met the Director. If she answers quickly, as though it's not very important, that's one thing. But I'll bet she'll be embarrassed and make up some story. You'll know."

"Why would she do that?"

"If I'm right, the first time she met Akenov was when he treated her for infertility. She would not want you to know that. The usual way IVF works is that the mother is treated with hormones so that she produces several eggs. These are collected and mixed with the father's sperm, and the ones that begin to divide are implanted in the mother's uterus. This is why twins and even triplets are so common with IVF."

Avalon paused, thinking, then went on. "Here's the important thing. Infertility can be caused either by the father or mother. If it's due to the father, a sperm donor is used, so that fertility clinics keep lots of anonymous sperm samples on hand, frozen in liquid nitrogen. But Akenov is in charge of these, and no one would know if he was using himself as a donor, would they?"

Marat shook his head and stood, looking down at Avalon. Then he turned away without saying anything and left her sitting there.

As she pondered her situation, Avalon knew that her most important task was to act as if she had no important task. As soon as she understood what Akenov was doing, she realized that she was part of his plan. The entire school, not just the clinic and research area, were his laboratory. She saw the entire setup very clearly—a laboratory with human subjects. She already began to suspect the nature of her role in the experiment and wanted no part of it. Inside the school and the Institute compound she was a prisoner. Outside she would be a fugitive, but a free fugitive. To get from inside to outside she needed help.

She looked out of the window at the open space between the dormitory, the school, and the institute. She watched students and a few adults come and go, and she was keenly aware that only adults went past the guard's kiosk and out the gate into the village. And for the first time, she considered the possibility that the compound must have rear entrances for truck deliveries of food and supplies. They would be locked of course.

The one person she could trust also happened to be one of the very few students who lived in the village—Marat. She found him sitting alone in the cafeteria, looking down at his plate, lost in thought. Avalon put a cup of vegetable soup on her tray, then walked over and sat down next to him. Marat didn't look up, but simply said quietly, "I can't believe it. I just can't believe he would do that."

Avalon said nothing for a few seconds. They sat like two people who had just learned of a death in the family. Finally Avalon said, "Just remember, you may be his son, but you are not him."

"I hate that . . ." He used a Kazakh words she didn't understand.

"Will you help me with something?"

Marat said nothing, but after a few seconds he nodded.

"I need to get out of here as soon as possible," Avalon said. "I want you to tell me all the ways in and out that you know."

In her mind's eyes she had already run through every escape she had ever seen in a cartoon or a movie, from catapults to tunneling to hiding in laundry trucks to wearing disguises. She could not hide in the village, so she would need time to put distance between herself and Akenov. During the day was out of the question, but even at night she would be conspicuous—a girl alone on the road. Early morning, a couple of hours before dawn might work. She could be out of the village, and then watch for a bus taking people to work in the city.

But Marat knew about one exit that she had not thought of. Like thousands of cities and towns in the former Soviet Union, the village received heat for all of its radiators from a central, coal-fired, water heating plant several miles away. Marat said that he had seen maintenance men in a basement under the cafeteria crawl into a tunnel that brought the steam pipes into the school and the Institute. He had never gone into the tunnel, but he had heard students say they had followed them for a long way under the village, maybe even farther.

By this time the cafeteria service people had disappeared into a back room for their own lunch, so it was easy for Marat to lead Avalon through the swinging doors into the kitchen, then through an inconspicuous door on the right that opened to a stairwell serving the basement storage area. He turned on the single hanging light and walked to the back of the room where he moved several large cardboard boxes to reveal a small door just big enough to accommodate someone willing to crawl through. The door was secured by an old black iron padlock, but Marat flicked it with one hand and laughed. "I know how to open these locks," he said. "Every kid does."

Avalon asked, "Can you unlock it for me, then leave it open?"

"Sure. But I should show you how to do it in case someone locks it again."

Avalon watched closely while Marat used the clip of a ballpoint pen to pick the lock, then left it hanging on the hasp. "Okay?"

Avalon nodded, then took his hand and led him closer to the dim light bulb so she could watch his face. "Marat, please don't get upset by this next question but I have to ask it. Can you trust your mother?"

Marat did not hesitate. "Of course! I know I can. Why?"

"Because someone else has to know about Akenov, just in case."

"I think she already knows."

"But how could she? You haven't told her yet, and you didn't even know until this morning."

"What I do know is that she hates Dr. Akenov, and now I know why."

"But she still works with him."

"She has to send me to the school here because that's where she works, but now I understand why she wants me at home, rather than being one of the boarding students."

"All right. I'll assume that she does know and leave it at that. Now I need you to draw a map of where I might be able to get out of the tunnel and then how to get into the city."

Marat looked over at the door, its open lock, then back at Avalon. "That won't be necessary. I'm coming with you."

Avalon let go of his hand. "No, you can't do that." She immediately wondered why she had said it. She had not planned her escape yet beyond the village other than losing herself in the city and trying to find Americans who could go with her to the consulate.

"Yes I can. I'll tell my mother and she'll agree. Besides, you'll be much safer with me. I speak Russian and Kazakh. I've been to the city. If anyone stops us you are my sister. I am part of your disguise."

She found herself smiling. In Kazakh she said, "сендер оң болып табыласыңдар."

"Yes, I am right. By the way, you speak Kazakh like my sister," he said. "If I had one."

Chapter 26

They agreed that Marat would find his way into the pipe tunnels from somewhere in the village and they would meet at the tunnel entry in the basement at 3 a.m. Avalon went to bed fully dressed and set her wristwatch alarm for 2:45. The tiny high pitched peeping woke Mariam, but she went back to sleep when Avalon whispered she was going to the bathroom. None of the doors at the school had locks, so it was easy for her to walk to the cafeteria, staying close to the walls out of the moonlight, then down in the basement. The lock was still hanging loose, and when she opened the tiny door she felt a wonderful sense of relief to see Marat crouching there, smiling. He handed her a small LED flashlight. "We're outta here," he whispered, then closed the door, hoping no one would notice the open lock for awhile.

They ducked, walked and crawled through the tunnel following the little white blue beams of their LED flashlights. "Just beyond the edge of the village," Marat said, "Then we can go up."

Avalon estimated that they crawled for almost an hour. They passed side tunnels, puddles of water that smelled like sewage. Their lights occasionally caught the fleeing rear end of a rat. Once they passed a small black and white cat crouched in a side tunnel, waiting for prey. It hissed viciously, but shrank back from their passage. With the exception of a few places where the floor of the tunnel leveled, they went down a slight slope. Avalon imagined she could feel the hundreds of people asleep above them. Twice they heard a car pass on a street. When Marat signaled a stop, her thighs and calves ached with a tiredness she had never known, and she knew that straightening her back would be painful.

They turned off their lights. Marat climbed up the metal bars in a masonry wall that served as a ladder for workmen. Marat had to bend his head down and pushed up on the heavy iron cover with his shoulders, moving it sideways with short, slow motions of his shoulders. When she climbed out, Avalon saw a dark starry sky and immediately recognize several constellations. More than once she had amazed Michelle and Don and friends by pointing them out in the clear summer skies of Oregon where stars appeared in such bright abundance they often confused even experienced amateurs.

Keeping carefully to the side of the road they continued down the slight slope toward the main highway that led west to the city of Almaty and east to the Narynkoll region on the border with China. Marat had memorized his mother's instructions about buses and streets in Almaty, and he carried a small map in his backpack. To confuse anyone trying to track their route, they took the first little bus east toward the tobacco and sheep herding area around the town of Chilik, but at the former collective farm village of Michurin they got off, crossed the road and waited for the bus to Almaty. The only driver who picked up passengers at the science village would say they had gone to Michurin.

In Almaty with dawn just extinguishing the streetlights, they melted into the crowd at the central bus station. From there they took trolley buses crowded with people going to work. At the corner of a Peace Street and Kabinbai Hero Street, they passed through an alley between the café on the corner and the adjoining old three story hotel. In the corner of the café building they climbed a few steps to the residential stairwell, and on the floor just above the café, Marat pushed three times on the bell button beside a heavy steel door. Almost before the third muffled ring finished inside, Marat aunt Makhabat, his mother's sister, opened the door. Her round brown face smiled broadly under her housewife's head scarf. She kissed them both.

A few minutes later Avalon and Marat were sitting on opposite sides of a small table in his aunt's kitchen. A clear plastic sheet covered the white enamel table top with its fading red roses. On the table between them from the spout of a ceramic teapot a thin wisp of steam twisted lazily upward and disappeared. They had finished their plates of dark brown buckwheat kasha and sour cream, and for dessert Makhabat had set down a plate of butter, a pile of toast and a jar of homemade raspberry jam. Marat was spreading butter and jam on the last two slices of bread and Avalon was drinking a third cup of very milky tea—Kazakh style—when the doorbell rang one long ring, then another.

Makhabat shouted down the narrow hallway in Kazakh, "Just a minute. I'm coming." To Marat and Avalon she said, "Put everything, dishes, everything in the refrigerator and go to the bedroom with your things. Just to be safe."

While they followed instructions she moved a few things in the hall to make noises of preparation, and when she heard the bedroom door close, she said from behind the security door, "Who is it?"

"Message," said a woman's voice. Better a woman than a man, Makhabat thought. She had a spy hole and looked through it before opening the door. That cheery voice had come from a sixtyish, very pale skinned Russian woman with worried eyes and a bun of black hair piled on top of her head.

Makhabat stood back and admitted the woman who said almost commandingly, "Close the door."

Makhabat closed the door.

"I think you should listen very carefully," the woman said in Russian, clasping her hands before her and looking over Makhabat's shoulder down the hall. "The message is this," she said and giggled nervously but without smiling. "First, the message is from the man who says he knows the girl whose picture is on the mantle."

"I don't have a mantle," laughed Makhabat.

"Please, never mind. Maybe someone you know has a mantle. It is from this man, and he says if you have guests now, they should leave quickly." The pale skinned woman looked Makhabat in the eye and saw her confusion.

"Well, you see, that's all I know," the woman said. "I am just doing what I was asked to do. You are old enough to remember the importance of secrets and when one did not want to know too much."

"You were asked to come by whom?" Makhabat said quietly, trying not to frighten the strange woman. "And can I ask who you are?"

"No you cannot," the woman said, and again she giggled as if she were playing a game and also found the world itself absurd. "I'm sorry. I have to go." She turned and opened the door and slipped out.

Makhabat called quietly after her, "Spacebo."

As soon as Avalon and Marat heard the door close with a turn of the dead bolt, they came out to the hallway. "I heard most of it," Marat said.

Makhabat repeated the Russian message and Marat translated.

Avalon said, "We have to leave. Do you have some ideas where we can go?"

Marat said, "Who sent the message and how do we know we can trust him?"

Avalon answered, "Better trust him than ignore him, because he knew about a photograph we have on the mantle at home. I can't tell you who it is because I don't know." She could piece together a few pieces—someone who knew they were missing and that a search was on, also knew about the picture on the mantle, or knew someone who knew. She was already moving to gather her bag.

Makhabat began to move also. "That woman was right—sometimes what we don't know keeps us safe. Sometimes." She spoke as she began gathering food from the kitchen to give them. Her grandfather had been sent to die in the Siberian forced labor camps of the Stalin era as an "enemy of the people." Her father had served a year in jail for participating in the 1986 Kazakh uprising against the Soviets. She made Marat turn his back to her, tucked a few packages of food into his book bag, turned him around and handed him a thick sheaf of paper money.

When Avalon and Marat stood in the hallway behind the door ready to be let out, she said, "Wait here. I'll go downstairs to take out the garbage and look around. She slipped out with a plastic bag of garbage and in a minute she came back without it. "Go now," she said. "Don't tell me where, but Marat, you know you have relatives here and in the east. I can honestly tell anyone who comes that I don't know where you went." She opened the door, kissed them both again and said, "Good luck."

Chapter 27

In the alley to Peace Street, Avalon asked Marat, "Can we flag a taxi?"

"Sure," he said, "But which way do we go? I need time to think. Maybe to your consulate."

"If they knew we were at you're aunt's, they know we are in the city and the consulate is one place they will wait for us." She had already understood that Akenov's plan would be a network of watchers on the move trying to pick up their trail. "First, let's get a cab and go into the busiest part of the city where they can't watch everyone."

They crossed Peace Street and stood in the gutter where Marat raised his hand in the signal that he wanted to hire a car. Many drivers were glad to participate in this system and earn the price of a meal or a liter of gas. Four or five cars passed before a battered green Soviet era Lada pulled to a stop. The man driving it was a middle aged Kazakh with down slanting sad eyes and a bow-like mouth. He leaned across the passenger seat and rolled down the window and asked in Kazakh, "Where to?"

Marat thought only a second and told him a destination across town. The driver said, "Two hundred fifty tenge."

In Kazakh Marat said, "Two hundred and maybe we go farther and pay more."

The driver smiled and reached a long arm over the passenger seat to open the back door on the curb side. Marat and Avalon slipped in. Avalon noticed that the windshield on her side was a spiderweb of cracks, and could see the driver looking at them in the rear view mirror. She knew what he saw—two young Kazakhs, maybe the youngest people who had ever flagged his car.

"How far to Keskelen?" Marat asked.

"You want to go to Keskelen?"

"My sister does," Marat said. "But I'm not sure. We'll talk about it."

The driver continued north down the gently sloping street until it leveled out and he turned left onto a broad six lane road. "Rayembek Avenue," he said. "If you decide to go to Keskelen."

When the driver turned his attention to the road, Marat told Avalon that Keskelen was a small town he knew about because the school's Kazakh teacher said that many people there spoke Turkish, a language akin to Kazakh. They threaded traffic for 20 min, the car rattling, the driver muttering, Marat and Avalon talking quietly in English. Avalon studied the road, trying to memorize it in case they had to come back this way. Suddenly Avalon said loudly in Kazakh, "Oh, stop here." They were in the middle lane and the driver swore, blew his horn and cut off another car as he swerved toward the side of the road. Marat was about to ask a question when Avalon kicked his foot.

Cars stopped behind them were blowing their horns. Marat handed the driver two bills and they jumped out onto the side of the road. The drivers of the cars that followed the old Lada back into the stream of traffic glared at them.

"Why here?" Marat asked.

"I saw a sign back there for a church."

"You want to go to a church?"

"The sign was part in English. It said, "Holy Trinity Catholic Cathedral.""

"You think it's American?"

"I'll explain as we go, but first we have to get across the street."

They began to walk back in the direction they had come, toward the sign, watching for a break in traffic. Avalon stepped out into the road to start crossing when traffic came to a temporary bumper to bumper jam, but Marat pulled her back. "They'll kill you," he said. They walked farther back up the road, and when he judged they could make it across an open gap between cars, he said, "Run, fast," and they did. They saw a blue sign with a cross and arrow pointing up a side street and followed it toward a tall sand colored stucco steeple. A broad stairway led up to wooden doors at the bottom of the bell tower. They looked at each other, Marat with a puzzled look on his face. Avalon finally said, "Well, let's go." The right hand door swung slowly open when she pushed on it. Marat followed her into the small vestibule, then through another pair of swinging doors where they stood looking down a long aisle that stretched to the altar. A life-sized Christ hung from a giant crucifix behind the altar, his arms nailed to the cross and spread wide as if to welcome visitors. The main body of the church was tall, brightly lit from its high windows and white stucco walls. The huge space was quiet, but in the silence they heard echoes from a faint murmur. Avalon whispered, "Look to the right of the altar—kneeling down."

A woman's head with a white cloth over it showed above the back of the front pew, her arms resting on the rail in front of her, her hands clasped. "She's a nun," Avalon whispered, then spelled it because she knew Marat might not understand. She motioned Marat to sit with her in the back pew. They waited. The nun continued to pray. After a few minutes Avalon said, "Wait here. I'll go talk to her."

When Avalon reached the end of the pew, the nun, still kneeling, turned to look at her. Avalon said quietly, "Excuse me." The nun stood up with difficulty. She was a tiny brown woman in a brown smock tied loosely around her waist with a white rope. A white headband was drawn across her broad forehead, emphasizing thick black eyebrows beneath it. She said something in Russian to Avalon and Avalon replied that she did not speak Russian, but English or Spanish. Sister Teresa's eyes widened and she said, "Yo soy Hermana Teresa a tus ordenes, mi hija. (I am Sister Teresa, at your service my daughter)."

In Spanish Avalon explained that she was an American, that she had come with a friend, a Kazakh student, and that she wanted to ask for help. The little nun looked toward the back of the church where Marat was sitting, then up at Avalon who was almost a head taller than her. The nun had her head tilted slightly to one side and a gentle smile on her mouth, her eyes calm and unblinking as she tried to look inside this Kazakh-looking girl who said she was American, who spoke no Russian but English and very good Spanish.

When Sister Teresa said, "Si?" the question was an invitation that Avalon accepted without hesitation. She began by telling the nun about attending the convention with her mother, the legal problems that suddenly emerged, her detention, and the school. Instead of saying what she knew about Dr. Akenov's work, she said, "I ran away because a man brought me a secret message from my mother that said I was in danger. I knew it was true."

Sister Teresa looked straight into Avalon's eyes, judging how much to believe of the strange account. Then she said, "I will call Father Carlos. He is an American. Please wait." Before Avalon could agree, the nun walked quickly to a door to the side of the altar terrace and disappeared.

Two minutes later the door opened and a priest paused at the threshold, looking first at Avalon, then at Marat. He wore the brown habit of the Franciscans, his tunic also tied with the rope of poverty. He was a short man, sturdy and broad with a closely shaved head. He strode forward and said in English, "I am Father Carlos." He looked to Marat in the back of the church and with a slow inward circling of his open hand invited him to come forward. Marat stopped by Avalon's side, and the priest seemed to take a mental picture of them together. Then he said, "Follow me please." He led them to a small room with a simple table that served as his desk where he unfolded three metal chairs and invited them to sit.

"I heard a rather strange story from Sister Teresa. You, Avalon, are an American and Marat is Kazakh, and you were both in a special school run by Dr. Akenov. Your mother sent you an email and said you were in danger."

"Not an email," Avalon said. She knew that *a message delivered by a messenger* would be less believable, but it was the truth and she said it anyway. She anticipated the priest's disbelief and added, "I don't have my passport or any other identification. But I am from Oregon and my mother is Dr. Michelle Murphy."

"Yes," Father Carlos said, "I have read about Dr. Murphy. She gave a lecture in Astana about the nature of consciousness."

"Yes, that's right," Avalon said, relieved that she had found a connection.

Father Carlos went on as if he had not heard her. "St. Francis, you know, spoke with birds and other animals, so perhaps he would have been interested in Dr. Murphy's lecture." The priest smiled and nodded toward Avalon. Her turn to speak.

"I came here with my mother. I was at the conference, but they would not let me go home because they said my adoption was not legal."

"Very few things in this country are legal," the priest said almost under his breath. "Who brought you the message and what is the danger?"

Marat looked at Avalon and she looked at him. The unspoken question was how much they should say. Avalon asked Marat in Kazakh, "Everything?"

Marat thought only a second and shrugged, "You say."

Avalon decided she could trust the priest. He listened intently as she told him what she knew about Arman Akenov and the school.

When Avalon had finished, Father Carlos sat for a while with his hands together in a sharp steeple just touching his lips. When he intertwined his fingers and lowered them to his waist, he said, "Thank you. We often hear that someone is 'playing God', but here is a man who really is playing God." He looked at Marat who was watching him, then back at Avalon. "Why did you come to us to ask for help?"

"I saw your sign on the street. It was the only one in English," Avalon said.

"I'm glad you saw it. We help those who need help."

"Could you help me send a message to my mom?"

"Of course. You can use the computer in the church office." He turned to Marat and asked, "What about you, Marat?"

"My mother knows I'm with Avalon, but she doesn't know where. I'd like to tell her we're okay."

"We'll talk about this while Avalon sends her message. Because your mother works for Dr. Akenov, we have to be careful." Turning back to Avalon, he said, "Perhaps no one reads our email, Avalon, but it would be best to say only those things you would not mind strangers reading. In fact, it may be best if you send the message to someone I know—my sister in California—and I will tell her how to send it to your mother." Then he smiled. "How is that for a suspicious old priest?"

Father Carlos took Avalon and Marat to the small church office where he sat down and slowly typed an email note on an old Samsung computer, addressing it to his sister. Then he rose and motioned Avalon to sit down and write below the note. "Be sure to add your mother's email address."

While Avalon wrote, Father Carlos spoke quietly with Marat in Russian after apologizing for having only elementary Kazakh. "Do you want to tell me anything that Avalon didn't say?"

Marat thought for a moment, then told him about the village and the school, and his mother's work with Akenov. After a few minutes Father Carlos leaned back in his chair and asked, "And would you like to know anything about me?"

The question surprised Marat. Almost without thinking he asked, "Why did you become a priest?"

Father Carlos nodded his head, smiling. "I could say that God called me. Do you believe in God?"

"I don't know," Marat replied. "Did you hear God?"

"No, not the way that I now hear you. God does not call the way a friend calls you on the telephone. God speaks to us in our own inner voice. In English we have a special word for this—faith." He said the word in English. "Do you know it?"

Marat shook his head.

"In Russian, I think we have only the word *belief*. Faith," he again said the word In English, "is to believe in what you cannot prove, what you cannot see, or hear, or touch, or feel, or taste. But you still know it to be true."

At this point Avalon stood up. "I sent the email to your sister. Do you think she will get it soon?"

Father Carlos chuckled. "Her computer is always on and she is almost always on the computer." When neither Avalon nor Marat spoke, Father Carlos continued, speaking in English. "Avalon, I believe I know what we can do to help you and your mother. You will be safe at the American consulate. Perhaps I can call and have them send someone to pick you up. And what of you, Marat? You can call your mother, of course."

Marat explained that someone had traced them to his aunt's apartment and that a woman had come to warn them to leave quickly. "I'm afraid she might be in trouble."

"What kind of trouble?"

Marat said, "I'm not sure, but she knew I was going to help Avalon, and she knows why."

"Was she upset? How did she feel about what you told her?"

"My mother is a very calm person," Marat said. "She wanted me away from the school."

Father Carlos shook his head. "If you two were not here, looking me in the eye, telling me this, I would not believe it."

"It's true," Marat said.

The priest smiled. "Yes, Marat, you say it is true. I believe you, but I have no evidence, no proof. In English we call that faith."

Avalon said, "But we can give you proof."

"I'm sure you can," the priest said softly. "Now I hope you can forgive me if I support my faith in you by seeking some proof. Let's talk about how to do that."

Chapter 28

Asel had thought of several ways she could explain Marat's absence from school. When she saw Dr. Akenov standing at her office door with an unreadable face, she chose the story that would win the most time for the fugitives.

"You know, then," she said without the usual morning greeting. "Marat was not at home this morning."

"And I'm told that Avalon was not in her room this morning. She seems to have disappeared," Akenov replied.

A picture of Marat stared at Asel from her desktop, the back of the frame toward Akenov. She looked into his eyes, then at Akenov. "I have to tell you that Marat said many unpleasant things about you last night." She was choosing her words very carefully. "He was very upset, but he said he could not tell me why he was so angry."

She was surprised by Akenov's odd smile when he said, "I'm sure they will be back soon." He said as if he knew something she did not know. "I will also call Professor Murphy and tell her. Maybe she will hear from Avalon." He gave Asel a list of tasks to do in the laboratory which did not surprise her. Akenov had never shown any interest in her personal life.

She said, "These can wait. I would rather take the day off and look for Marat. I might be able to find both of them."

"Yes, of course," Akenov said. "But these lab tests are very important and will take only few minutes. Just quick scans of tissue cultures and change media if cell growth has affected pH."

She agreed to do them, and Akenov went away. She put on her lab coat and scanned the cultures as quickly as she could, but it took more than a few minutes because three of the Falcon flasks needed attention. When she returned to the office she hung up her lab coat, picked up her bag and left. Dr. Akenov's office door was closed as she went by. As soon as she found a cab in the village, she reached in her bag for her cell phone. It wasn't there. She was sure it had been there, today of all days because she expected a call from her sister or from Marat. She thought quickly about when she might have used the phone and forgotten it. But someone must taken it, because the only moment her bag had been separated from her was when Dr. Akenov had sent her to the lab. Akenov. He could see who she had called, or anyone who was calling her. He would see last night's call to her sister.

She asked the driver if he had a cell phone. He did, but he said she would have to pay for the call. She handed him several bills, and he said sincerely, "Thank you," and handed her the phone. She called a woman she did not know well but who lived not far from her sister. The woman lived on a small pension from a printing house and spent her time drying flowers and making fanciful and highly colored pictures with their parts. Adventurous enough, Asel had decided. She was correct. The woman told her that she had delivered the message to Makhabat's apartment. Asel erased the call record before giving the phone back to the driver.

Asel still had the day to herself but had no idea where Marat and Avalon would be. She knew the chance of finding them walking along a public road was very small, so she asked the driver to return to the school where she at least had access to a telephone. When she passed Akenov's door on the way to her office, it was still closed. She had a strange feeling that she did not really work here. Everything was familiar, but she was no longer part of it. She sat at her desk staring at a picture of Marat. "Where are you?" she asked quietly. She remembered her missing cell phone, and without hope, she looked again at every shelf and windowsill where it could be. Nothing. She was sure that Marat would call that phone as soon as he could, and whoever had the phone would know where he was calling from.

When the phone rang Arman Akenov did not answer. Smiling a small but real smile, he wrote down the number that showed on the screen, and he waited for the caller to leave a voice message. Twenty minutes later a beautifully polished, black Nissan Pajero stopped at the curb directly in front of the church steps. Despite the heat of a late August day, the driver and the two Kazakh men who stepped out of the car wore black leather jackets over black shirts. They also wore black slacks. One of the men standing was huge. The other was average height but seemed small by comparison, yet moved with an athletic grace. They climbed the stairs to the church doors with deliberate speed, paused, and the smaller man knocked loudly on the door. After a few seconds when the door had not opened, he pulled on the handle and the tall door swung out to reveal the vestibule and beyond it the long aisle to the altar and beyond the altar high on the wall the crucified Christ. A very tall monk with a shaved head and long flowing brown tunic swept down the aisle towards them, smiling.

In Russian he said, "The church is always open for you. I am father Woytek. How can I help you gentlemen?" He did not offer his hand, only a welcoming smile of great confidence and self-assurance.

The smaller Kazakh said, "You have two students from Good Shepherd School who must go back with us. They should not be away from their classes." Both the speaker and the larger man seemed ready to walk deeper into the church, following their already searching eyes, but Father Woytek did not move out of their way.

The priest said, "Good Shepherd—yes I have heard of this interesting school. But two of its students we do not have here. As far as I know, no students from Good Shepherd attend our church. And I should know because I direct the youth programs."

The larger man said, "We'll take a look."

"I see that you will," the priest said. "I will be glad to accompany you to the church, our Sunday school and youth Hall, and our offices. I think you will understand if I cannot invite you to visit the monastery or the convent." He turned and crossed himself before he began to walk slowly down the aisle. The two Kazakhs followed, their eyes scanning the empty pews for anyone hiding.

When they reached the front of the church the little man stepped boldly onto the altar platform and looked behind the altar and into the choir stalls. The larger man opened the door to a small room behind the choir stalls and came back out in just a few seconds. The smaller man said to the monk, "Now your offices."

After the two men had inspected the church offices and found nothing, they walked around the church building, past the Mary shrine and shrubbery. The little man said, "We know they're here, so we we'll search your living quarters."

The three of them were standing in front of identical buildings behind the church, one for the sisters and one for the brothers.

"I am afraid that is not possible," Father Woytek said. He smiled apologetically. "To tell a lie in our religion is a serious sin. I can tell you in all truth that you will find no students in these buildings."

The little man said sharply, "I'm afraid you have just sinned. We'll look anyway."

The priest smiled down on him and said almost apologetically, "And I know I have not sinned."

"Do you? Stop talking nonsense. Let's have a look."

"Perhaps there is something I don't know," Father Woytek said, "but I have never heard of school officials who have the right to search anywhere, no less a place of worship and a monastery."

The smaller Kazakh said, "There are many things you don't know. Some things maybe you don't want to find out. Do you understand?"

The priest looked back at him without speaking for a few seconds, looked at the larger man who was just an inch or two shorter than himself. One thing the priest understood was that confronting men in black leather jackets and black pants who traveled in black SUVs could be very destructive.

The priest made an exaggerated gesture of looking heavenward before he said, "So it must be that you shall have your way. Give me but one minute to be sure we do not embarrass anyone." He turned quickly, knocked on each door in turn and gave short instructions to the person who answered.

The smaller Kazakh made a phone call and listened intently, then looked at incoming images on his screen and smiled.

The big Kazakh was visibly surprised when the tiny sister Teresa gave him a short nod and motioned him to follow her into the convent. Father Woytek invited the smaller man to follow him, but the smaller man said, "Wait, Ivan. We're done here. Priest, do you have a cell phone or computer where I can send you a picture of the girl?"

Father Voytek thought for a moment, pulled a piece of paper and a pen from his pocket and printed in block letters his email address.

The Kazakh man worked on his phone for a minute. "I just sent you a picture of the boy and another of the girl." He paused in a way that Father Voytek thought was almost playful. "Just in case they do come here or you see them somewhere."

Father Woytek said, "Now you have seen our simple life and we have seen something of yours. Perhaps you will return as our guests. Or perhaps you will invite us to your mosque."

"We're not interested in fairy tales," the small man said, and the big man laughed.

The priest made a short bow of goodbye, then watched the unwelcome visitor walk rapidly out of the church. He heard them slam the car doors behind them, and then the Pajero's tires squealed as it accelerated away from the curb. In a flimsy metal garage a few blocks from the church, Father Carlos was sitting on a cushion in order to see through the window of the church's excursion van. Sitting beside him in the passenger seat was a tall nun with beautiful, sad eyes and a few wisps of blond hair visible beneath her cap. Sister Nina had joined the church in Russia to serve as a nursing sister, but decided to spend her life in Almaty because of the fluent Kazakh she learned from her mother.

Marat and Avalon sat in the last row of four rows of seats hidden by the smoked windows. They were removing the matching brown tunics and rope belts that had disguised them when they left the church grounds as Akenov's men were talking to father Voytek. Father Carlos' cell phone rang. When he was done listening he turned around and said to Avalon and Marat, "Father Voytek says we can return now and decide what to do."

They entered the back door of the church. Avalon noted that several monks and nuns appeared to be walking in slow meditation outside the church fence. She was sure they were on a kind of guard duty, a comforting confirmation that the monks and nuns were now on their side. Father Carlos, Sister Nina, Marat and Avalon crowded into Father Voytek's small office and closed the door behind them. Avalon was glad to brush the hood back from her head, untie her rope belt and shed the too warm tunic.

"We have some time, Father Voytek said, "but not much. They sent me these pictures from Dr. Akenov so that we could recognize you if you came here." From the screen Marat stared back at them from a classroom in the school. A second picture showed the side of his face as he read a book. Father Voytek clicked the mouse and Avalon's face came on the screen with the white Tien Shan mountains and a Shepherd's yurt behind her. Avalon gave a short laugh at seeing herself. "What? That's impossible!" Father Voytek clicked again and a picture of Avalon standing between Nurzhan and Mariam, an arm over the shoulders of each. The picture in fact was a video and almost as soon as it came on Avalon said in Kazakh, "I'm glad to meet you."

Avalon laughed again and said, "Very clever." The others in the room turned to look at her. "That's not me," she said. "I have never seen that yurt, and I never stood with my arms around both of my roommates, certainly not when being introduced to someone."

"Photoshop," said father Voytek.

"No," Avalon said. "I think I have just been introduced to my twin sister. Now I need to go back to meet her."

"Not to the school," Father Carlos objected. Father Voytek scratched his fuzzy head. Sister Nina pursed her lips.

"Look at the pictures again," Avalon said.

Father Voytek brought up the picture by the yurt, then the brief video of Avalon—or whoever it was—saying, "I'm glad to meet you."

Everyone stared at the girl on the screen as Father Voytek ran the image and video several times. He and the others looked back and forth between Avalon and the girl on the screen.

Father Carlos said, "If she's real, I agree that she must be your twin."

"I should have known," Avalon said. "Good Shepherd specializes in twins. Now I've got to call my mom."

Chapter 29

August, 2020

The "DNA Suite" ring tone woke Michelle at 3:30 in the morning. Avalon's voice was clear and Michelle had never heard her so excited and apparently happy. She couldn't keep up with the words she was hearing. "Slow down Avalon! You woke me up, and you're talking too fast. What is this great news?" Although she had so far heard only the happiness and excitement in Avalon's voice, she felt her own hopes rising like a tide turned from outgoing to incoming. This growing elation was unreasonable, she knew. *Hope*, she told herself, *is an emotion. Think.* But she could not stop it. Avalon's voice was so clear, so alive that she seemed already home again. Michelle reached to the bedside table, turned on the light, picked up the pad and pen and began to take notes. Several times she had to stop Avalon and ask her to repeat part of the story or fill in details.

A half hour later when the call had ended Michelle was wide awake. She had several pages of notes and could begin to think. She understood the gist of Avalon's call, but nevertheless went through her notes and numbered each burst in the order it should have in a coherent report. She knew from years of experimental science that raw data, the bare notes of events and observations, might tell an accurate story, but focused thought and concentration over time almost always discovered the questions hidden in the answers, important gaps that the mind's thirst for story had either missed altogether or filled in with convenient fiction. For instance, who were those men in black SUVs and black jackets who had appeared at the church? She began to list other questions:

Where had Avalon's twin sister been for 13 years?
Had Akenov known about the twin sister all along? And if so, why had he never said anything?
What else did Akenov know about Avalon's birth family?
How would Akenov use Avalon and her sister?
How would the discovery of her identical twin change Avalon's view of her situation? Would it alter her attachment to Michelle and America?

Michelle was happy that Avalon had now found not only an equal but a twin sister. Although circumstances had changed quickly, the controlling fact had not: Avalon was in the hands of a very dangerous man whose intentions Michelle understood, but without knowing the consequences for Avalon.

She thought about calling Don despite the early hour, but decided she should use the few hours before he would be up and about to organize her thoughts and think about her options. To calm the thoughts buzzing in her head like an awakened swarm of bees, she slipped out of bed and sat on her yoga mat, legs crossed, spine straight, hands cupped at her navel, eyes almost closed. Little by little the buzzing of thoughts faded into the faintly audible beat of her heart and the slow whisper of her breathing. Then the "DNA Suite" sounded again on her cell phone. Without uncrossing her legs she leaned toward the beside table, picked up the phone and with her back again straight and her eyes almost closed and her thoughts very still, she answered.

"Dr. Arman Akenov is troubling you," the voice said. Michelle did not know that this greeting was a traditional apology in that part of the world for phoning unexpectedly.

"In fact, you were troubling me very much even before this call," she replied, but very coolly.

"I have called to help you. Believe me."

"Tell me why I should believe you."

"Only a few minutes ago you talked to Avalon. That is good, is it not?" Akenov asked the question with obvious delight as if asking credit for having delivered an irresistible and exotic sweet.

"Dr. Akenov," Michelle began formally but he cut her off.

"Please—call me Arman."

She could think of a lot of things she wanted to call him, but Arman was not one of them. "Dr. Akenov," Michelle repeated. "You know and I know that you are holding my daughter against her will. I also . . ." She had been about to say that she knew his lab and his work on animals and crops was a front, but Lipkovich had warned her not to risk putting Akenov on the defensive because his defense might quickly become an offense. She took a deep breath, closed her eyes, and said, "I am also willing to cooperate with you in any reasonable way to get her back. So let's get down to brass tacks. What do you want?"

Akenov laughed. "You have brass tacks. Do I need brass tacks?"

Of course, he didn't understand the old American phrase, and come to think about it, she knew the meaning but not why brass tacks meant blunt facts. "Forget the tacks," she said. "It's an American idiom. Facts—that's what it means. I repeat, tell me straight out what do you want?"

Damn, she had used another idiom—straight out. "What can I do for you?" she concluded in what she hoped was a conciliatory voice.

"Yes," Akenov replied, "straight out or maybe straight in I will tell you. Is very simple. You are world expert on genetics and development of nervous system. I am expert on improving genetics by in vitro fertilization and cloning. Avalon has mutation that produces extreme intelligence, so we are both interested in her. I propose that we work together to understand what her mutation does during embryonic development, and then share it with all people of world."

Michelle was startled by Akenov's proposal and took a moment to think. "And if I refuse?"

"Ah. I don't think you will do that. Avalon is here. You are there. Avalon's twin sister is also here. She is, I know for fact, identical twin sister, identical genes. So, you see we have—what is it called—back up copy, no? So, do I need Avalon? Perhaps not. Who I do need is you. You could change my mind, I'm sure, but I think twins belong together."

Michelle was about to say, *You only need one of them*, when she realized that what she really had to deal with was that Avalon needed her sister, almost certainly the only other person in the world like her. That thought grew instantly into several more, like a mental fractal. Avalon had as much as said she could not leave her sister. Michelle knew the gift she could give Avalon was a life with this sister. She also knew that she wanted not just Avalon but both of them.

Michelle said to Akenov as pleasantly as she could, "Yes, Arman, you are right. They belong together."

Chapter 30

Before Sister Nina took her to the church office with an international phone line, Avalon said, "We don't have very long before those men return."

"How do you know this?" Father Carlos asked.

"I watched from the window and saw that their car did not go toward the main road. Dr. Akenov wants them to make sure I return."

Sister Nina stared at her. "Do you know these people so well? Do you Marat?"

Marat shrugged. Avalon replied, "I don't know them well, but I know what fits. If we don't leave, they will come here again. If we leave, they will follow us."

While Avalon called her mother, Father Voytek, Father Carlos and Sister Nina held a quick meeting in Father Voytek's office. They could hear Avalon excitedly talking to her mother in the next room, then she was back with them in just a few minutes. They tried to persuade her not to return to the school, but she was adamant. They proposed to call the consulate and ask them to send a car for her, but Avalon insisted that she would not go to the consulate without her sister. "Dr. Akenov would certainly not let my sister leave the school. Furthermore, the Kazakhstanis say that I am not an American citizen because I was adopted illegally, so I'm not sure what the consulate could do anyway."

Avalon looked around and saw the brown robed monks and nuns staring at her. She understood their confusion at confronting a girl her age who spoke with such confidence. Father Voytek took the lead. "Let's say we take you back to the school, but we also inform the consulate about the situation and what we are doing."

Before she could answer, Father Carlos said, "I propose that Sister Nina and I will drive you back and we will stay there with you." Several others murmured approval, and Avalon nodded. Father Voytek rose from his chair and stood in front of Avalon and Marat, looking down at them with a smile. He made the sign of the cross, and said, "Bless you my children. May God keep you safe."

Sister Nina wanted to be with them to verify Avalon's assertion that many of the students at Good Shepherd were in fact Akenov's children. She said that if she became convinced, Father Carlos would begin to apply careful pressure on Dr. Arman Akenov. The first goal was to secure the safety of Avalon and Marat, and the second was to force Akenov to abandon his project. None of them suspected how vast Akenov's ambitions were, or what he was capable of doing to protect his goal of making the Kazakh people into the next human species, a superior species that would inherit the Earth.

As they pulled out onto Raimbek Boulevard, the main road, Avalon looked back and said, "We're being followed."

Father Carlos looked in the mirror and saw the black SUV turning out of the small street on to Raimbek. "Since we are going to the same place," he said, "there will be no chase scene in this movie." He continued to drive slowly and carefully.

Father Carlos parked the van near the gate to the compound that held Good Shepherd School and the institute. He met Sister Nina as she came out of the passenger side door, and the two of them stepped into the archway that housed the guard's kiosk. Beyond the kiosk at the entry to the schoolyard a sturdy iron grill on a steel track blocked the passage, but a smaller grill could be opened for visitors.

Inside the kiosk a beefy Kazakh guard sat at the counter with the guest book between himself and the window. He was talking on a cell phone, paused when he saw them, then nodded his head and said a few more words into the phone before putting it down. He looked down through the window at the two visitors in their brown tunics, as if they were aliens, but aliens arriving on a predicted schedule. He was staring at Father Carlos when Sister Nina surprised him with her flawless Kazakh, asking to see Dr. Akenov. The guard was one of those Kazakh men who like to boast that they had to use a riding whip on his wife just once, then let it hang on the wall. As soon as his surprise faded, he said dismissively, "The Doctor is not here. Who are you?"

Sister Nina did not answer the guard's question, but stated, "Then we will speak with the rector of Good Shepherd." The guard shoved the registry book through the window onto the ledge before them. When they had signed in, he studied their names. They had both signed in Russian characters. "Wait," he said as he picked up the telephone on his desk, dialed three numbers and said into the phone, "They wanted to see Dr. Akenov. Now they want to see you." He listened for a few seconds, then spoke to Father Carlos. "Wait. Someone will come for you." He pressed a button and the smaller iron portal swung open. Sister Nina walked past the kiosk into the courtyard. The guard commanded, "Wait!" She paid no attention and walked slowly into the middle of the courtyard, looking around. "The woman is a fool," the guard said to Father Carlos as he would have to any man from whom he expected understanding.

Father Carlos shrugged. "We are only men," he said. "She is a woman."

"Stubborn women need to be whipped," the guard said. He liked the way that sounded.

Another woman, this one in an all white uniform worn in hospitals, came out of the lab building and crossed the courtyard to greet them. "Good morning. I am Asel Satpayev. How can I help you?"

Father Carlos had not expected a woman, but he managed to show no surprise. He studied her a moment and said with a smile, "Yes. I am glad to meet you. Very glad to meet you."

Asel glanced into the kiosk where the guard stood listening, then turned to Father Carlos. "The rector asked me to find out what you want."

"It's about some missing students. I'm sure Dr. Akenov knows."

Asel looked at him for a few second, surprised by such a confident, deep voice coming from a small man dressed in a monk's robe and sandals. She had never before talked to a Christian monk or even seen one. "Perhaps he does," she said. "Please follow me." Father Carlos followed Asel and heard the gate clang shut behind them. As soon as they had walked to where Sister Nina was waiting, Asel asked, "Where are Avalon and Marat?"

Then she looked over the shoulders of the little priest and saw two men in the black uniform of the Good Shepherds standing at the gate. She used a key to open the door and said, "We should go inside."

They passed Nurlan's office where he sat reading a newspaper and drinking tea. He looked up and nodded at Asel and stared at the two Franciscans trailing her. In Asel's office Sister Nina sat in one of the chairs Asel offered, but Father Carlos stood behind his, hands on its back. He asked Asel, "May we speak privately here?"

"Yes, but very quickly. Dr. Akenov will want to know where Avalon and Marat are. He's expecting them."

Sister Nina replied. "The children are in the van parked outside. Avalon insisted on coming back."

Asel was startled by this news, and lowered her voice. "Avalon should not be here at all."

Sister Nina looked at Father Carlos. The priest said, "She knows that. We know that. But she would not agree to come with us except to the school. Once she found out her twin sister was here, she wouldn't go anywhere else."

Asel stopped suddenly just at the door to the medical laboratory complex. "What twin sister?"

Father Carlos explained about the phone call and Avalon's sister as quickly as he could. Asel shook her head in disbelief. "I have not seen any such girl." Asel's lips tightened. "Be that as it may, the children should not be here. Marat is my son, and he should be at home. Avalon shouldn't be here."

"Why do you say that?" Father Carlos asked. "You can tell us, I hope."

"I don't have much to tell you. Only that she received a message that she is in some danger here. And of course, I know that her mother must want her back. I mean her American mother."

"And the danger? What is the danger?"

"The danger can only be from Dr. Akenov."

"Yes," Father Carlos said very slowly. "I believe we know more than you do, and I will tell you, but this is not the time or place."

Asel nodded. "In that case I will take you to see him now."

When they stepped into the corridor they saw Dr. Akenov standing in front of his own door, feet apart, his arms clasped behind his back and a look of pure contentment on his face. "Welcome to our school. There was no need to hide in Asel's office. My shepherds told me you had come."

Sister Nina walked a few paces down the hall to stand in front of Akenov and stared intently down at him. "I understand there are many young children here. May I see them?"

"Of course, if you wish. They will enjoy meeting such an unusual visitor," Akenov replied. "But first, there are two unusual children I wish to see. My shepherds will bring them to us and Avalon will meet her sister. This will be a happy occasion for all of us."

Chapter 31

Shortly after Michelle's unpleasant phone conversation with Akenov, he sent an email with an attachment. The text read:

> To: My Respected Colleague, Dr. Michelle Murphy
> Together we can solve all problems and even change world for better. I attach proposal for our joint work. With your help, I am certain we can succeed.
> Do not worry about Miss Avalon. I assure you she is very secure here at Good Shepherd.
> Be healthy and happy,
> Arman

Michelle read the text again. He had typed it himself. A translator would have used the articles "a" and "the". Except for that kind of carry-over from Russian that has no "a" or "the", his English was very competent. And surely he knew the difference between "secure" and "safe." She downloaded the attachment and opened it, surprised that Akenov had chosen to write it as a letter, rather than a standard research proposal. But then she realized that he had probably never in his life had to write a proposal to a government funding agency, as she and other American scientists must do in order to support teams of graduate students and post-doctoral associates in their labs. She was bitterly amused when she began to read.

My dear Michelle,

I hope you will forgive the trouble I am causing you. I know my English is not perfect, so to be as clear as possible I am having my words edited by Nurlan, who has worked for me as a translator for years. After reading this note, I think you will agree that the trouble is for a good cause, perhaps the most important scientific experiment ever undertaken. I will begin by summarizing what we both know. By the way, Nurlan just made me smile by suggesting an English phrase. I want to be sure we are on the same page.

As you know from your visit, I have developed cloning techniques that improved Kazakh horses to the point that they are now winning races in international competition. As you also know, I am the director of the Good Shepherd School where I am bringing together Kazakh children with unusual intelligence, thinking that someday the methods we use on our horses might also be applied to humans. However, humans are not horses, and this is a project that has already taken ten years of my life. With another ten years we might see some results. I am not that patient!

Like all scientists, you and I devote our lives to improving the human condition. Avalon has given us an unparalleled opportunity to do so. She carries in her genome a unique DNA sequence that increases human intelligence immensely. It is, I believe, a major step in evolution. This is a treasure beyond measure and must be preserved for all humanity. But Avalon is Kazakh, so like our mineral and oil wealth, this gene is property of the Kazakh people. Our work begins here!

How can we do that, you and I working together? Now I'm afraid you will scoff, perhaps even be shocked, but I have considered cloning as one possibility. My research with horses taught me how to achieve nearly perfect results, but you may rest easy, it will not happen. This is not a proper breeding program!

But there is another way to achieve our goal, and I think you have already guessed what it is. Again you will be shocked, but I am fully aware of the progress you have made at TransTek, just as your NSA knows the content of private calls made by the German Chancellor, Angela Merkel, and probably our president as well. I know that you have succeeded in developing a viral vector that can transmit Avalon's FOX gene into other human beings.

I too have some experience with these viral vectors. I believe that you and I, in a collaborative effort, can do this. I will provide young Kazakh children for the experiment, and Avalon's gene now that she is here. I have a fully equipped laboratory, so you will provide the technical expertise to design and fabricate the viral vector.

Of course, the details need to be worked out when you arrive. I have reserved a first class ticket for you and hope that you will soon see Almaty again, and Avalon. She has a surprise for you.

Your friend and colleague,

Arman

Toward the end Michelle's face reddened, and she stood up, knocking her office chair backwards. "That son-of-a-bitch!" Michelle surprised herself as she almost shouted at the computer screen. She looked up and verified with relief that she had closed her office door. She never used this language, but she had never felt this kind of rage. She stood up and looked for something to tear apart. She imagined throwing the laptop through the window, but she immediately realized that would be "shooting the messenger". Her rage began to focus instead on the subject of the message, a man she would have gladly pushed out the window. That pompous gnome with the phony smile, Dr. Arman Akenov.

She sat at her desk and closed her eyes, propped her elbows on the big calendar, and held her head in her hands, scanning Akenov's proposal again. She was looking for nuances, hidden meanings, and confirming the clues that revealed what he must already know, what he had already achieved, and the true intent and results of their collaboration—if she chose to collaborate. Akenov was well along in his experiment, and his proposal was feasible. She had to tell someone. Don, of course, but he could only listen. There was one other person who knew enough about Akenov to think about possible solutions, perhaps even had the resources to actively help. She picked up her land line phone and dialed a number she had been given to memorize, then listened to an odd series of tones that ended with a click and a voice she immediately recognized.

"Yes?"

"I just received a new email. From Akenov."

"Is your computer on?"

"Yes."

"Hang up and open Google."

Michelle replaced the phone in its cradle and used her mouse to click on the Google Chrome browser button. The usual screen appeared, but in a few seconds it faded and she was surprised to see it replaced by the face of Evan Lipkovich.

Michelle couldn't help laughing. "How did you do that?"

Lipkovich smiled. "Haven't you been reading the newspapers?"

"You mean about the NSA penetrating Google and Yahoo? Does the CIA do that too?"

Evan continued to smile, but shrugged his shoulders.

Michelle nodded. "OK, I get it. But how did you get into my computer?"

"It's not hard. Any computer linked to the Internet has an IP address, and we have an encrypted version of Skype to let us talk to other agents. And now you. However, one thing our friends at the NSA can't do is to intercept CIA conversations, so don't worry about that. What did you need to talk about?"

"I know what Akenov is doing, and why he is keeping Avalon at his school."

"I'm listening."

"It's not what we thought. A viral weapon is not his primary interest."

"Then why did he order the SV40?"

"I think he was just experimenting with ways to deliver genes, but if he had hacked the TransTek computers, he knows that I'm working on a more efficient delivery system. I'll get to that, but first let me fill you in with what Avalon observed at the school. You do know that he's running a school for gifted Kazakh children?"

"Yes."

"And that he also directs a fertility clinic?"

"Yes."

"What if I told you that many of the children at the school are twins, even a set of triplets, many more than would be expected statistically in a normal human population."

Michelle paused to let Lipkovich think. Finally he said, "So, for some reason he's choosing gifted twins for the school? Twin studies—not uncommon if I remember my psych classes. But I don't see a connection."

"Avalon did, but she had a few more dots to connect than just the twins. She realized that an unusual number of the children shared several genetic traits, as though they all had the same parent."

Michelle watched Evan's image on the screen. After a moment he shrugged. "Still don't get it."

"Do you have a photograph of Akenov?"

"Sure." A few seconds later Evan's face was replaced by what looked like an unposed close up of Akenov, who seemed to be looking down at something as though he were reading.

"I guess I shouldn't ask how you got that photo."

"That's right, you shouldn't."

"Anyway, take a look at his hairline, and tell me what you see."

After a moment, Evan said, "It's a widow's peak. My dad had one, and I do too."

"That's right. Now look at the left ear that shows in the photo. Would you say the lobe is attached or hanging free?"

Michelle could not see Evan while he looked at Akenov's image, but heard his voice say, "It's attached."

"Right again. Now put the photo away so I can see you."

When Akenov's face was replaced by Evan's, he looked puzzled. "I don't see where you're going with this."

"Suppose I told you those are traits that can be passed from parents to their children, like you and your dad. Avalon noticed not just the unusual number of twins at the school, but also that they shared a whole set of traits like widow's peaks and attached ear lobes. In fact, nearly half the kids had the same traits, not just the twins."

Michelle paused, waiting for Evan to make the connection. It took a couple of seconds. "Damn! Is Akenov sowing wild oats, as my dad used to say?"

"Not just sowing them. He's running a breeding operation, something like his horses, but with humans. He thinks Avalon's genome could be used to improve the Kazakh gene line. He has even considered cloning her. She's his prize. You know why?"

Lipkovich paused. "Because she's super smart maybe because she's your daughter?"

"Because she has an identical twin."

After Michelle told him what she knew, he thought for a moment, then said, "Maybe that's good. He doesn't need Avalon."

"That's exactly why he does need her." When Lipkovich did not respond immediately, Michelle continued. "Think! Avalon is nurture and her twin sister is nature—got it? And now he's got what scientists like to have, a control copy for whatever he does next. He's known about it since they were born. He put Avalon in the orphanage and left her sister with the parents. I can't prove it, but I'm sure when we came to adopt Avalon, he made it possible so that she would have a very different upbringing. He knew who we were. That's why the adoption suddenly got easier in the last couple of days and the red tape disappeared."

Lipkovich was making notes she could see that by the way he looked down.

"Even the Kazakh student in Don Koskin's lab might be involved. Do you remember talking to him?"

"Sure. His name was Bakhyt Esemberdiev."

"Did you know he dropped out? He told Don he missed his family and was going home. I don't believe it. He came to Don's lab with a reference from Akenov."

Lipkovich paused for a few seconds, as though he was considering what Michelle had said. "Yes, that could be. Can you prove it?"

"Not in a court of law. Do I have to prove it to you when my daughters being held hostage?"

Lipkovich's voice was friendly and reassuring. "No you don't. You don't have to prove it to me, but I may have to prove it to somebody else if we want to get Avalon out."

Michelle sighed. "Not just Avalon, she won't leave without her sister. You have to understand she's not like you and me or anybody else in this world. She's been starving for somebody like herself, and now she has not just somebody, but a twin sister."

Seconds passed while Michelle and Evan looked at each other on their computer screens. Then Evan said, "Why would he tell you that? And how?"

"Simple. An hour ago, Akenov called me out of the blue. Without actually saying so, he made me understand that he was the one who had arranged for Avalon to be detained. She's basically a prisoner at his school, and he's using that to force me to go back. And of course now Avalon doesn't want to come home without her sister. He knows that. He said he would let them both go but only after I had agreed to help him with his research. And it's pretty clear he wants to create a Kazakh transhuman."

"What did you tell him?"

"I don't intend to help with his research, but I didn't say that. I am considering a trip to Almaty to discuss his proposal, buy some time. He already bought me a ticket. At least I can be with Avalon, and maybe figure out a way to get her out of there."

Michelle paused, looking directly at Lipkovich on the screen. "You're going to help me, aren't you?"

Lipkovich leaned back in his chair, looking away from the screen for a moment, then turned his attention back to Michelle.

"One way I can help right now is to warn you that Akenov is a sociopath."

Michelle laughed. "I know that."

"What you don't know is that he's a murderous sociopath. There are stories about him going all the way back to childhood, and he didn't get to where he is by being a nice guy. Here's the deal. You have never before met someone like Akenov. Academic politics can get rough, but at least there are no corpses. If you do go back to be with Avalon, keep one thing in mind. If he gets what he wants, he has no reason to keep the two of you alive. In fact, you will be a threat to him because you know too much and he has enemies who would love to know what you just told me. Give him what he wants, and he will arrange for you and Avalon and her sister to disappear."

Michelle was chilled by what Evan was saying, and could not respond. He was right. She had never considered that she and Avalon might be in danger.

Evan continued, "Regarding the CIA helping you to get Avalon out of there, it's like a chess game. I have a few pieces in play, but first we need to consider the range of what can be done. One possibility is to do nothing and let it play out. Another is to mount an extraction operation, but we can't risk that with all the children at the school. We have some friends in the Kazakh government, but that will take time, and we just can't go in and tell the president's men that the country's most famous scientist is a sociopath, a Kazakh Hitler. We can try to begin talks with Akenov, but again that takes time and right now he's holding the trump cards. Michelle, I must tell you that the CIA can only respond to specific threats to national security. If what you told me about Akenov is correct, his activities are not a clear threat, so I can't provide direct help. It's not our job to rescue US citizens who are being held by a foreign government. That's the job of the State Department who can bring diplomatic pressure to bear. So with your permission, the first thing I will do is to bring Rhonda Grable into the picture and see what she can do."

"Meanwhile, Avalon is stuck in Almaty with a madman. Her sister too."

"But if your scenario is right, he wants them to stay healthy."

Another emotion swept through Michelle that she had never before experienced, a sense of cold despair mixed with her hot rage toward Akenov. She felt powerless, and a representative of the most powerful nation on Earth had just told her he could not help. She reached over and turned off her computer.

That evening, Michelle sat back in the deep comfort of her favorite soft leather armchair in Koskin's living room, sipping the dark purple hibiscus tea they kept just for her. Don, Jodie and Julie were on a couch, looking at her expectantly, like students.

"Well," she began, "I have some news about Avalon, and now I need your advice."

For more than an hour she led them through the complex situation that had engulfed her life and Avalon's. Her three friends were astonished by the revelation that a crazy but ingenius Kazakh scientist was running a human breeding operation and had essentially kidnapped Avalon, or more precisely, Avalon's unique genome with its FOXP5 sequence. Jodie was particularly outraged. "He wants to clone Avalon? Can he really do that?"

"No one has succeeded in human cloning, to my knowledge," Michelle replied, "but not for lack of trying I'll bet. And someone will succeed, soon."

Don said, "It would be the dream of every monarch, chief or dictator. Isn't that why so many conquerors from the Greeks to Genghis Khan took captive women for themselves and their sons? They weren't making exact copies, but they sure spread their genes. I can imagine someone like Kim Jong-il wanting to have a copy of himself growing up as a member of his family. Instead he got Kim Jong-un."

"A narcissist's dream," Michelle said. "Akenov has apparently developed some specialized techniques from his research on horses. He has considered trying those on Avalon and I suppose on her twin."

"Haven't most countries banned that kind of human experimentation?" Don asked.

"There's even a UN Declaration condemning it," Michelle replied, "but it's still going to be done by someone. Cloning isn't technologically difficult, but it's hit and miss. Basically all you need to do is to suck the nucleus out of a skin cell, pop it into an egg cell that has been enucleated, let it develop into an embryo and implant the embryo in a uterus. Remember Dolly the sheep, the first successful mammalian clone in 1996? That's exactly how it was done, but it took 277 eggs of which 29 got to the embryo stage and were implanted. Three lambs were born, but only Dolly survived. So there, as I see it, is Akenov's biggest problem if he wants to clone Avalon and her sister, and he knows it. Given those odds, where is he going to find 50 or so women who will donate eggs, and another 50 who are willing to have embryos implanted in their uterus? A few infants might be born, but there might be just one Dolly for all that effort."

Michelle was listening to herself and growing impatient, but she knew Don and Jodie were listening, taking it in. They wanted to help. Don said, "If I had to guess, I'd say you're pretty sure these obstacles can be overcome and that maybe Akenov will try it."

"Yes. In 2008, the Stenmegan Corporation in La Jolla produced five human embryos from male human skin cells and donated eggs. Of course, this was simply an experiment to see if it was possible, and they destroyed the embryos. But the Japanese have now cloned clones of clones of clones. In mice, of course."

There was a moment of silence while Michelle's audience of three pondered all that she had told them. Then Don cleared his throat. "You said that the lousy odds were one of Akenov's problems. Are there other methods?"

Michelle smiled. "There are, and that might be a way for me to get Avalon home. Think about it. He's a breeder, and if you want to breed horses, he needs stallions and mares, right? And if he wants to breed a race of advanced humans, he's going to need both boys and girls. A clone is always the same sex as the donor nucleus, so even if he manages to clone Avalon he will end up with only females. It's a dead end."

"Unless," Don said, "those females could then pass on their FOXP5 to their children."

"Akenov would have to live to a very old age to see that," Michelle said. "He's not a patient man."

Jodie said, "So how does that help you?"

"From my work with TransTek, I know how to do what Akenov wants. Did I tell you the TransTek computers have been hacked? He must know that the techniques we are developing will produce both males and females, and all of them will have Avalon's FOXP5 gene. No cloning process is needed, and the first children will be born within a year."

Don unwrapped his arm from Jodie's shoulders and leaned toward Michelle, elbows on his knees. "If someone other than you said that, I would write them off as a crank and show them the door. Can you explain?"

"I have a non-disclosure agreement with the company, so I'm probably liable for damages if it gets out that I'm telling you about this. But I think I must, because I need your advice on whether to go ahead."

Michelle looked around at her three friends, and saw that she had their absolute attention. "I can use TransTek proprietary discoveries to produce a viral vector containing FOXP5. It will be targeted for cell surface receptors on human eggs, and it will be infective, like a cold virus, so it can be delivered by inhalation. Any woman who is infected comes down with symptoms resembling a common cold while the vector inserts FOXP5 into her 300,000 eggs. After she recovers, any child she bears will have Avalon's gene."

The other three looked at each other, and began imaging the consequences of what he or she had heard. Julie said, "Am I right—a woman catches a cold from your virus and any children thereafter are as brilliant as Avalon?"

Michelle said with very measured words, "They will all have the gene that makes Avalon unique, similar to the mutated gene that makes Homo sapiens different than all the previous homo varieties."

Jodie reacted faster than the scientists, Julie and Don, who were arranging lists of consequences and questions in their minds. "Michelle, that's crazy. You're not God. You just can't wave your wand and change the whole human race."

Don and Julie looked to Michelle for her answer.

"I'm flying to Almaty next week to negotiate with Akenov," Michelle said. "He wants me to work with him, but I'll tell him that I already completed his project. Tell me why I shouldn't propose to trade the vector for Avalon. He'll have what he wants and we can go home."

Don sat back shaking his head and crossed his arms in front of him. "Michelle—I know you, and I know what you would be giving Akenov. And I know that if this virus escaped, it would permanently change the human germ line. Maybe for the good, but we don't know that. We don't know all the consequences. Even Avalon, wonderful as she is, has only just become an adolescent. We have no idea what this gene might do when an adolescent becomes and adult. Do we?

Michelle did not hesitate. "No, we do not. Now let me ask you a question. Do you think we should hide what we know? Even we keep it a secret, someone else will find out sooner or later. And if they find out about FOXP5, I can't image they won't use it."

Don sat back in his chair and crossed his arm. "I agree. We're already putting human glial cells in other animal brains and thinking about beefing up the supply in human brains. Silver and Wray at Duke discovered that HARE 5 enhanced cortical development, and that's public knowledge now.

Jodie said, "Wait a minute—glial cells? HARE5?"

Michelle said almost impatiently, "In 2007 they reported that glial cells from a human brain make mice smarter. HARE 5 is an enhancer gene whose human variant seems to account for why we have larger brains than chimps and apes." Then she looked at each of them briefly, returned to Jodie and asked, "Do you think we shouldn't play god?"

Jodie said, "Isn't it called hubris? The gods used to punish us for it. Read your Greek tragedies."

Don said, "It's not a Greek myth any more. We really are discovering godly powers, a lot faster than most people imagine."

"Here's my thinking," Michelle said. "Let Akenov have some form of this gene to play with. It's more important to have Avalon and her twin safe, because they're the going to show us what FOXP5 does to actual human beings."

Julie got it. "I see what you mean. Knowing what we know, we're already years ahead of everyone else."

"We are, but only if the girls are with us. I've got to have something that will let me bargain with Akenov."

Don took a deep breath, and when he spoke, he was unusually firm. "You asked for advice? My advice is to not even think about it. Akenov could play his games with your vector, but what if it escaped? You would be releasing an infective virus that could do… who knows what? Mental disorder, growth problems, maybe cancer! I'm sure there are other ways to get Avalon back. We must be missing something."

Jodie put a hand on Don's shoulder. He leaned back in his chair, realizing that he was giving orders, not advice.

Michelle waited for questions and Julie spoke. "You and Don and Julie are miles beyond my understanding, but I have to say that this as an enormous gamble. No one can think of all the possible consequences. Like in ecology, where you can't make just one change without affecting everything else?"

Michelle nodded. "I take your point. Of course, TransTek is very concerned about potential problems like that, because they need FDA approval for any drug they develop. All the management and scientists at TransTek attend seminars on bioethics, and we have monthly meetings to review old questions and new issues. We have reviewed this viral vector and its potential intensively. The viral vector is called a lentivirus. It serves well because it has enough room in its interior to hold the gene we want to deliver. To make delivery precise, we have engineered it in three ways. First, we make it like a cold virus so that a small number inhaled can reproduce in the respiratory tract and make millions more copies that are released into the blood. Second, we introduced a targeting molecule on its surface that can be designed to fit cell membrane receptors in specific body tissues."

Michelle turned to Don. "I do understand your concern, but we do know a lot about how certain viruses can activate cancer. There is absolutely nothing in the vector that could do this. All it does is deliver a gene."

Don was obviously and uncharacteristically impatient. Michelle seemed to have become someone he didn't know. "Michelle, this isn't you talking. You're not listening either. It's not the virus—it's the consequences of the FOXP5 gene over a lifetime. No one has had a lifetime to observe what its consequences could be. In bioethics it's called the precautionary principle, and prudent vigilance. In other words, don't jump before you know where you might land."

Jodie took a turn. "However small a chance, it could escape, right? And it could infect every woman who catches a cold and change her forever. It might change the human race forever."

Michelle stood up as if shaking them all off her back. "Look, I get the point. I understand, but those kinds of changes happen all the time. It's called evolution. Don, you were at my talk about Ma, weren't you? Avalon is Ma 200,000 years later. If Ma hadn't existed, we might still be living in trees or competing with chimps and gorillas. Knowing what you know, if you could have helped Ma survive so that her mutation would make humanity possible, you would have done it, wouldn't you?"

Don was shaking his head. "I don't buy that. If evolution were a scientist, we could say it experiments on very small groups or even individuals. If the trait gave the new variety an advantage, then evolution's experiment continues to ever larger populations. Granted that mutations occur all the time, and some of them are positive selective factors that make evolution possible, but you are proposing to experiment on the entire human population at once." He paused, then continued, "You and Akenov. Whether you like it or not, you'll be teaming up with a sociopath."

Michelle opened her mouth to speak, but Don held up a hand and continued, hoping he might bring Michelle back to reality. "And you and Akenov would do this—change billions of people's genetic makeup without their permission. Even if this were a population of mice or fish or termites—the experiment on an entire population would be ecologically unethical. Even if you were sure everyone would be smarter and have no negative results—you have to give people a choice."

Don stood and walked to the front door, then turned and said, "Michelle, I understand that you are a mother with a child in danger. I can't tell you how sorry I am that all this has blown up. But there must be another way to rescue Avalon. I need to get outside, take a walk."

When the door closed, Michelle looked over to Jodie and Julie. "How about you two? Am I a mad scientist? Am I playing god?"

Jodie shrugged, but Julie laughed. "Michelle, Don's right, but if you need a volunteer to test your vector let me know. I'd love to have a baby like Avalon."

Michelle smiled at her. "You'd be a good mother, but not yet. And when it becomes this personal, I see Don's point about experimenting on other people. I couldn't do it to you."

They sat back and sipped their coffee and began to relax, ready for whatever came next.

The front door opened and Don backed in, shaking his umbrella on the stoop. "Forget the walk. The drizzle is turning into a downpour with lightning and thunder. Listen, Michelle. We've been friends for a long time, but if you make that vector, I'm reporting you to the TransTek people, to the FBI, the CIA, anyone I can think of."

Michelle looked up at Don and shook her head. "That won't be necessary. Talking to the three of you has given me a better idea. I can make a fake vector to trade for Avalon. It will do everything it should, but the gene it delivers will be an ordinary human gene, FOX P2. Akenov won't know that until the first children are born, and by then Avalon and I will be long gone."

Michelle stood up, looking around at her friends. No one said anything, but she could see growing understanding in their eyes. Don finally nodded.

Michelle walked to the door and opened her umbrella. "Now, if you will excuse me, I need to get to work."

Chapter 32

Hours later, rain still poured out of the skies and ran in rivers down the streets when Michelle walked a few blocks to the Riverfront Research Park where TransTek maintained its headquarters and laboratories adjacent to the university campus. As a member of the company's scientific advisory board, Michelle had certain privileges. Because university rules prohibited commercial research on campus, the company had provided her with lab space and the equipment necessary for producing transgenic organisms, particularly the viral vectors that would be used to deliver desired genes to targeted tissues. She had a team of two Ph.D. scientists and two technical staff to work with. Her team was currently developing a virus to carry the gene for a protein called Factor 8, which was missing in patients with hemophilia.

Michelle scheduled her hours around Avalon's time at home as much as she could, which meant very irregular appearances—a few hours early some mornings, afternoons when Avalon had an after-school activity, some evenings when Avalon helped a friend study or visited with Don and Jodie. The guard at the TransTek gate was not surprised when she showed up and held her photo ID for him to examine. Even though she had been coming to the lab for several years and was a familiar figure, protocol demanded that the guard take her ID in hand along with her driver's license and check the photo against her face. He handed it back and said, "Another late night Dr. Murphy?"

Michelle smiled up at him. "Yes, I'm afraid so Charlie. Can't have my pet cells starving."

He laughed and waved her on. Michelle drove slowly through several turns in a maze of buildings, parking in her reserved space behind 239. She cracked the car door just enough to open her umbrella, then grabbed her briefcase and stepped out into the driving rain. Rather than going through the front door of 239 where she would need to be ID'd by another guard, she used her building key to open a side door that led directly into her lab.

Her research staff had gone home and the lab was dark. She pushed a glowing button that turned on the lights and walked over to a tissue culture incubator filled with several hundred rectangular screw-topped containers made of a clear plastic. After pulling on latex gloves, she removed one flask from the bottom shelf. It was half filled with a pink fluid, warm because the incubator was kept at 37 °C, the temperature of the human body. She sat down at a microscope, turned it on and placed the flask on the stage. The microscope was inverted so that she could look up through the clear bottom and focus on the cells growing there.

Under her breath, she whispered, "Good." These were human embryonic cells, maintained in a frozen state and made available by the National Institutes of Health to researchers who could justify their need, such as TransTek. The human body is built from approximately 200 different tissues, each with its own set of surface receptors and antigens. TransTek had ordered virtually all of the cultures, and this one had been originally derived from the ovarian tissues of a 6 month old female fetus whose mother had been killed in a motorcycle accident. The cells had grown and reproduced over the past 3 days, and now filled the bottom of the flask in a single layer. They were confluent, every cell touching surrounding cells, which caused them to stop growing. She could use them next day to test the viral vector.

Michelle marked the flask and replaced it in the incubator, noting its position in her iPad notebook. She went on to her next task, which was to prepare a viral vector. She had told Don and the others that she would make a fake vector with the human FOX P2 gene. She realized she had said that in large part out of her embarrassment of having even thought about releasing a new gene into the entire world population. Between that conversation and arriving at TransTek, she conceived what she thought of as a compromise. She had decided to prepare the actual FOXP5 gene in a second vector, but targeted for cerebral neurons instead of ovarian tissue. She knew Akenov would be suspicious. If he challenged her, she could show him the FOXP5 sequence in the second vector and demonstrate that it would infect cells in culture. The genes were relatively simple, just 864 nucleotides. They were identical except for two mutations in FOXP5. She would need only a few hours to make several micrograms of each. Referring to her iPad notes from 5 years ago, she looked up the sequences in the database and saved them on a thumb drive, then pasted the sequences into the computer that controlled a DNA synthesizer.

At the university, if she needed a specific gene, she would purchase it from one of the commercial companies specializing in DNA synthesis. This was cheap, a dollar per base, and saved time. But at TransTek the research was proprietary and had to be kept secret. That meant that DNA synthesizers were available in every laboratory. Synthesizing such long molecules had been unthinkable just a few years ago, but the new synthesizers could produce DNA with specific base sequences at a rate of one base per minute. Even at that rate, the FOX genes would not be completed until sometime the next day, so after she saw the machine turn on and begin the process, Michelle began the next step in her procedure, which was to prepare a virus as a carrier.

Michelle walked to the next room where the company technicians maintained virus samples in baths of liquid nitrogen, each equivalent to a large thermos bottle. She again consulted her notebook, then went to a barrel-sized thermos marked 7C. After drawing protective mittens on over her latex gloves, she pulled out a rack of small tubes. White frost crystals immediately blossomed on the rack, and drops of liquid nitrogen splashed from the rack onto the floor where they boiled into a mist that drifted across the room. She carefully removed a small tube labeled 7C-L3, the code for one of the proprietary lentivirus strains that could be used to deliver a desired gene to mammalian cells. This strain was one of her key inventions for TransTek. Its coat had been modified with receptors from an ordinary rhinovirus that causes colds. It was very new technology, and as she worked she reviewed in her head the briefing she would someday present to the company CEO and managerial staff. She was used to speaking to classes with hundreds of students in the audience, at scientific meetings like the recent one in Almaty, but also to high level management of companies she advised. Like most scientists, she could write lectures in her head, only later transferring them to Powerpoint or Keynote slides for the presentation.

The words played in Michele's mind as she worked, almost like a script. "Gentlemen, I am pleased to tell you that we have succeeded in establishing a protocol that will put TransTek far ahead of the competition in gene delivery. Today I will tell you about recent developments in my research group that will allow targeted genes to be delivered by inhalation, rather than injection. You might think this is impossible, but just consider the rabies virus that travels through the blood stream until it binds to the surfaces of cerebral neurons and is taken up by endocytosis. The virus begins to reproduce, finally killing the cell and causing the disease. By the way, it doesn't take the bite of an infected animal to transfer rabies. Years ago, when I used to explore caves in New Mexico, I learned that one of my fellow explorers had contracted rabies by inhaling cave dust containing the dried fecal matter of rabid bats. I'm sorry to say that he died, because he didn't know he had been exposed until the symptoms began and then it was too late for a vaccine to help. I knew nothing about genes back then, but when I began to work with TransTek I remembered that a virus can deliver its genes to specific cells in the body simply by inhalation. What we have done is to engineer a virus that can deliver a gene by that route, but not cause disease in the process.

"We have managed to incorporate the receptors of a common cold virus into the surface coat of a lentivirus, the preferred vector for gene delivery. I'm sure everyone in this room has discovered for themselves how infective a rhinovirus is, and suffered the upper respiratory tract symptoms. We designed our viral vector to infect the upper respiratory tract as usual, but then to enter the blood stream and travel to the target tissue. The TransTek target is the liver sinusoidal tissue that normally produces Factor 8, an essential blood clotting protein. In hemophilia patients, that gene is missing and they must inject themselves with factor 8 every few days throughout their lives or they could bleed to death from a minor injury. The cost of the factor 8 injections is approximately $100,000 annually. Imagine how important it will be for those patients to be given a true cure for their disease, simply by inhaling a dose of TransTek's product. I will leave the financial details to Mr. Burkhart, your CFO, but I think everyone here realizes what profits are possible for the first company that can bring targeted genes to market in a spray bottle."

The virus container was thawed, and Michelle placed it in a sterile glove box. Although the virus itself was not infective, normal procedure still required protecting the laboratory against accidental spills. Having clear and invariable routines in labs that worked with both dangerous and benign organisms created life saving habits. So the glove box was completely closed except when samples and reagents were being inserted through an airlock. Children seeing a glove box for the first time are amused, because the gloves reach out to them like the arms and hands of an invisible denizen of the box. To use the box, Michelle inserted her fingers in the inverted gloves and pushed them into the inner space where she could manipulate the virus samples and reagents. This part of the procedure was done according to a standard protocol called recombinant DNA technology invented 40 years ago.

Michelle carried out the first step, which was to add a gentle detergent reagent that would release the viral nucleic acid content. This would take an overnight incubation, so she set the container in an inconspicuous rack near the back of the box. Next morning she would use the CRISPR/Cas9 method to prepare the nucleic acid and insert the FOXP2 and FOXP5 genes. Finally she would reassemble the virus in its infective form and test its efficacy on the tissue culture she had checked earlier. If it was a successful transfective agent, the FOX genes in the chromosomes would be labeled with a fluorescent complementary probe and appear as a fluorescent spot.

But a tissue culture alone was not likely to convince Akenov. Michelle needed to show that it worked in a human being, and there was only one woman available. Despite Julie Flanagan's enthusiasm to become a volunteer. on the drive home from the lab Michelle decided to test the viral vector on herself.

Chapter 33

"All the world's a stage," Michelle quoted to herself as she boarded the first flight out of Eugene's local airport. "And all the men and women merely players." She remembered her Irish grandfather quoting the English Shakespeare, then merrily adding a line from the Irish playwright Sean O'Casey, "And most of us are desperately unrehearsed." She also remembered the first time Avalon heard those lines. She had asked Michelle, "So what is my role, and how do I rehearse?"

Michelle had answered, "Your role is to be who you are."

"And who is that?" Avalon asked

As Michelle took her seat she said to herself, *I think we're going to find out very soon.* A man took the seat next to her and immediately settled in to read the Wall Street Journal along with the scotch offered by the flight attendant in first class. She couldn't help thinking of another man sitting next to her 13 years ago on a flight to Almaty, a man called Hank.

Michelle's flight took her by phases into her new role on the stage of her life, a short hop with a stop in Seattle, then eight hours across Canada and the Atlantic to Amsterdam, followed by a six hours layover in the international lounge. Half of the passengers on the Air Astana flight out of Amsterdam were Kazakhs returning home from business, shopping, visiting relatives, or vacationing.

Eight hours later Michelle walked off the plane in Almaty at four in the morning, almost twenty-seven hours and three airports behind her. She was finally able to stretch. Even for her small frame, the Airbus 320 had felt very cramped, particularly with a fat Kazakh businessman who needed a bath sitting next to her. She stretched luxuriously at the bottom of the stairs, then joined other sleep dulled passengers shouldering or towing their carry-on luggage toward the waiting bus. The bus had no seats, so everyone stood and tried not to trip over the luggage. Michelle was once again jammed against her obese seatmate, who made no effort to give way. In fact, she sensed he subtly pressed forward when she tried to gain a little distance. She tried not to inhale his odor of vintage sweat dried in his clothes and new sweat from his body.

She groaned when she saw the long line at Passports and Customs. The agents were notoriously slow, checking and rechecking passports and visas, opening luggage and pawing through the contents as though every passenger was smuggling drugs into Kazakhstan. Feeling disheveled and smelly, worn out from the hours she had spent in planes to get here, she resigned herself to another hour in line before she could finally get to her hotel.

Then Michelle saw something unusual. Two agents were slowly working their way up the line with flashlights, ignoring the men but looking carefully at the women. Suddenly her fatigue was forgotten. She recognized one of them as the woman who had questioned her at the airport a month ago before leading Avalon away. The woman caught Michelle's eye and immediately walked over, ignoring the other passengers.

"Professor Murphy? Can I see you for a moment?"

Michelle realized that the polite question was more than that. It was an order. She mumbled under her breath, "I don't think I have a choice." The woman ignored her. As she followed the agents through the customs line reserved for flight crew, she was at least thankful to be spared the long wait in line. It was as if the play she was in had started early. They took her to a small room to the left of passport control, and when they entered she realized that this was where she had lost Avalon, apparently to the vagaries of international adoptions. She now knew that it had all been planned.

"Please be seated." When the woman began to speak, Michelle had to drag herself back to the present. Through a fog of fatigue, she tried to concentrate.

"Professor Murphy, my name is Batima, and I am truly sorry for the misunderstanding about your daughter's adoption. We have discovered that the error had to do with someone who was entering data into our computer system." She shook her head. "That person is no longer with us."

"That's great," Michelle said. "So I can take her home now?"

"Yes, that is possible if there are no further complications. I understand that she has been staying at Good Shepherd school? She was lucky. It is one of our best schools." This woman certainly knew her part in the script.

"Can I go to her now?"

"Of course. A car is waiting for you. However, this is customs, and as a formality I must check your carry-on bag. Do you mind opening it for me?"

The male agent lifted the bag onto the desk while Batima pulled on a pair of latex gloves. Michelle found the zipper pull and opened her bag. It was neatly packed with the usual things a woman takes for a brief trip. Batima carefully lifted out the clothing and placed it next to the bag, then removed and opened a small toiletry case. She inspected each item, opening a makeup kit containing the usual lipstick, mascara and a small perfume spray dispenser. She sprayed some on her wrist and smelled it.

"Very nice Professor. I like the aroma. But why are you smiling?"

"I'm sorry, I couldn't help it. You are being so careful."

Batima eyed Michelle doubtfully, then picked up the next item, a glass apparatus with a small rubber bulb. "What is this?"

"It's called an atomizer. I have hay fever that sometimes becomes so bad I can hardly breathe. That lets me spray the medicine into my lungs."

Batima, seemingly satisfied with that explanation, replaced the items and closed the kit. Only a book and a pair of high heeled pumps remained, so she picked these out and felt inside, finding that each of them had bubble wrap packed in the toe.

Michelle said, "Please be careful. There are glass vials inside."

"Can you undo the wrapping for me?"

"Of course, but I will need a knife for the tape."

Batima said something in rapid Kazakh to her colleague, who nodded and pulled a bunch of keys out of his pocket. There was a small pen knife on the chain which he opened and passed to Michelle. While she was cutting the tape, Batima asked, "Why did you hide those in your shoes?"

Michelle glanced up at her. "I didn't hide them. It just seemed like the safest place." She handed the knife back to the agent and carefully unrolled the bubble wrap on the desk to reveal two small glass vials. They did not have caps, but instead were sealed by melting shut the glass at the top to keep out air. Each had a thinner neck which could be broken to open the vials.

"What are they?" Batima asked.

"They are scientific gifts for Dr. Akenov, to thank him for taking such good care of my daughter."

Batima looked skeptical, but then glanced up at the male agent who gave a barely perceptible nod. Without another word, Batima turned and left the room. The agent watched impassively as Michelle packed her belongings again. After she zipped up her bag, he walked over to a second doorway and held it open. Even though it was just 6 a.m., with early morning sunlight showing through the windows, the noise of the concourse immediately flooded the room and Michelle was surprised to see that her fat seat mate was apparently waiting for her. He took her bag and led the way to the exit. Michelle looked back for the agent, but he had already disappeared into the crowd.

A large black Mercedes was parked at the curb with its engine running, ignored by a policeman who was professionally directing other traffic around it. When the trunk popped open and her former seat mate carefully placed her bag inside, Michelle realized that he was actually going to be her driver, and that her entire trip had been orchestrated since she got on the plane in Amsterdam. The driver opened the back door of the Mercedes and waited for her to get in, then settled himself behind the wheel. Michelle heard a click as the doors locked, and became aware once again of the intense unwashed odor of the man. Michelle could hardly catch her breath and needed to open a window. As the car left the airport and turned into a surprisingly busy avenue filled with traffic, Michelle asked the driver, "Do you speak English?" The man did not answer. Instead he pointedly turned on the radio and found a station that played the sort of wailing plaintive music that permeates Middle Eastern and Central Asian culture, a female voice accompanied by some sort of scratchy stringed instrument. He pushed another control and the air conditioning came on. Michelle sighed then settled back into her seat, watched the unfamiliar scenery for a few minutes, then closed her eyes and fell asleep.

A half hour later, Michelle awakened when the car jerked to a stop. For a moment she couldn't recall where she was. She sat up and saw that they were on a narrow street shaded by giant poplar trees. Across the sidewalk she could see a large gate with metal bars set in a two story concrete wall. It reminded her of a prison she had once visited in Portland to collect a blood sample from an inmate who had volunteered to be a subject for a TransTek research project. Fully awake now, she waited. If they had brought her to Avalon, she was content to let the play go their way for a while. The driver sat behind the wheel, looking straight ahead, saying nothing. He left the engine running with the air conditioning on, but now lit a cigarette, the acrid smoke almost choking Michelle. She tried to open her door but it was still locked. She mustered a steely voice of command and said, "Put out that cigarette—please!" It worked. The driver snubbed the cigarette in the ash tray but said nothing, didn't turn to look at her or look at her in the mirror. She did not like the new feeling the silence brought—that she was alone and powerless, under the control of forces she didn't understand. She leaned back in her seat, now shivering not just from the chilled air.

A few minutes later she was startled by the loud click of the door locks releasing and the trunk popping open. She immediately got out, gasping when the summertime heat hit her but gratefully filling her lungs with fresh air. She saw that a smaller gate within the main portal now stood open. At the same time a hand pulled down the trunk lid and revealed behind it a small woman, shorter than Michelle, in what looked like a lab coat. The woman held Michelle's bag. The Mercedes immediately accelerated away, leaving the two women standing there, looking at each other. The other woman smiled and said, "Dr. Murphy? I am Asel. Welcome to Good Shepherd."

Michelle sized her up and saw no bad intention, but she was quick to assert herself. "Where is Avalon?"

Asel's smile faded. She stared at Michelle for a few seconds, then nodded and said, "I am a mother too, and understand your concern."

Asel paused, as though thinking what to say next, then continued, "Avalon is here."

Asel turned and led Michelle through the gate, nodding to the gatekeeper, and Michelle heard it clang shut behind her. *A prison*, she thought to herself. Not a school.

"Dr. Akenov will see you this afternoon, but I'm sure you will want to rest and freshen up. You will be in one of our guest apartments. They are mostly for parents of our children, but with school in session we have few visitors."

"And Avalon?" Michelle said politely.

"I understand," Asel said with a voice that seemed not just sincere but sympathetic. "I think patience is best."

The apartment was one of four on the second floor, and Michelle saw that it was Spartan, with minimal furniture—a double bed, chest of drawers and desk. A television perched on the chest. She looked for a telephone but found nothing. At least she had a separate bathroom and small shower, which was what she really wanted.

Asel was standing at the door, and said, "I will come back after lunch at 1400. Is there anything else?"

Michelle was beyond trying to think what else she might need. So she just shook her head and stepped out of her shoes, a hint that one thing she needed was privacy. Asel nodded, then backed out and quietly closed the door.

The shower was heavenly, although the water never got hotter than luke-warm. Michelle had brought her own shampoo and conditioner, and the pleasant citron aroma washed away some of her fatigue along with the grime of travel. After drying herself slowly, she slipped under the sheet and blanket. Within seconds she was asleep.

A gentle knock at the door woke her. As she swam up out of deep sleep, she stretched hard, but gasped when sharp cramps knotted both her calves. She threw back the covers and stood up, knowing that if she could work the muscle, the cramps would release the knots in her muscles. At a second knock she yelled, "Hold on! Give me ten minutes please!" The knocking stopped, and she heard footsteps hurry away. With the cramps fading, she hobbled over to her bag and laid out the clothes she had decided to wear for the meeting with Akenov, a light grey pants suit and pearl blouse with an open neck. After getting into fresh underwear she brushed her teeth and took two aspirin. She decided not to bother with makeup, but did wash her face with cold water, hoping to get some color into it. With 5 min left, she walked to the bedroom and slipped into her pants suit. Sitting on the bed, she carefully removed the plastic wrapped vials from the toes of the shoes, placed them in a small hand bag and slipped her feet into the heels. Nearly ready, she returned to the bathroom mirror to put on a thin gold necklace, stud earrings and a quick brush of her sleep tussled hair. She glanced at her watch. Ten minutes had passed, so she walked to the apartment door and opened it to find Asel waiting there.

"Oh! Why didn't you knock?"

Asel smiled. "I was afraid to."

Michelle laughed. "I'm sorry I shouted. It's just that I was so tired."

"Believe me, I do understand. Did you get a little rest?"

"Ten minutes ago I was fast asleep. It's a good thing you came by, or I would have slept right through to morning. But I could really use some coffee if you have any."

"We don't have coffee here, but a pot of tea is waiting in Dr. Akenov's office."

"Will Avalon be there?"

Asel looked away, hesitated, then said, "I don't think so. Dr. Akenov wants to speak with you first."

Michelle caught the tone in Asel's voice. "Is something wrong?"

"No, nothing that I know of. Come. We're a little late."

Asel led Michelle down stairs and out into the central quadrangle. Michelle looked around as they walked across, taking in the tall and massive poplar trees, the walls with their strange symbols, a play area with the usual swings and slides. It was quiet. "Where are the children?"

"At this time of day they are finishing their last classes. Then they have several hours free before dinner." Asel pointed to glass doors in the opposite corner. "You can see the dining hall over there."

They had come to what looked like another wall, and Asel used a key to unlock a steel door, then held it open for Michelle to enter. She led the way up a flight of stairs where they came to a carpeted hall with art on the walls illuminated by soft light from hidden ceiling lamps. It reminded Michelle of the business decor of the president's office at her university. She had often thought that administrators loved comfortable surroundings because their jobs were so tedious. She glanced at the paintings as they walked by. They were all rural landscapes, a variety of settings, but every one had a flock of sheep and a shepherd somewhere.

Asel stopped before one of the doors. Michelle saw no nameplate or other identification. She knocked gently, then opened it a crack. "Dr. Akenov, Professor Murphy is here to see you."

"Yes, come in, come in!" Michelle shivered at the hearty tone, and thought to herself, *Said the spider to the fly.* Asel pushed the door open, then stood aside. Michelle took a step inside and paused to look around just as Avalon had a month earlier. Akenov was arranging papers at his desk, but rose as high as his short stature allowed and came to Michelle. Although he had been at his desk working, he wore an immaculate white lab coat with the Institute's logo embroidered on in national blue and gold on the left breast pocket, and above it in Kazakh two words that had enough familiar letters that Michelle could guess would be his name.

Dr. Akenov placed his hand on his chest, which happened to be over the logo and his name, in the Kazakh ritual greeting, then held it out to shake western style. Michelle hoped he did not notice her brief hesitation. *All the world's a stage*, she told herself, and she took his hand firmly but briefly. It was a dry small hand and she wished she had the strength to crush it. She let it go and felt soiled.

"Professor Murphy. Welcome back to Almaty. I hope your travel wasn't too tiring. Asel has prepared tea for us."

Asel was still standing at the door, but took the hint and followed him to the suite of upholstered chairs and couches around a low table, where she began to pour the tea from an elegant deep blue and bright gold teapot.

"Please, Professor Murphy, have a seat."

Michelle sat down across from Akenov.

"How do you like your tea?" Asel asked.

"Black is fine."

After the tea was poured, an uncomfortable silence followed. Michelle looked down at the tea service, unable to think where to begin. Then she noticed a beautiful gold object next to the service tray, a sculpted running horse being used as a paperweight. She couldn't help staring at it, a tremendous shock running through her mind when she realized it was identical to the artifact that Hank had shown to be a fake antiquity shortly before his death. Avalon had mentioned it in her first email to Michelle after her detention.

Akenov sipped from his cup and said, "You like it? This ornament discovered in tomb of Khazakh king, maybe queen, twenty years ago."

Michelle tried to recover, thinking she would ask Avalon about it later. "It's beautiful." Then she looked directly at Akenov. "Where is Avalon."

Akenov said matter of factly, "She is here, of course."

"Then of course, I want to see her."

"Soon. She is remarkable girl, and since she is subject we will discuss, better just you and I." He turned to Asel and said, "Professor Murphy and I must have private conversation now." He waited until Asel had carefully closed the door with a light click, then turned back to Michelle. He saw that she was staring at him, appraising, unsmiling. He smiled at her, but waited. Now that Michelle was confronting Akenov, she was uncertain how to proceed. How much should she reveal of what she knew? She had not rehearsed this moment. She decided to feel her way until she knew more of Akenov's intentions.

Akenov continued to smile at her without speaking. She had tried to take the lead with her question, but he dominated with silence. Unlike the direct and often blunt Michelle who had never even manipulated a boyfriend, Akenov had a lifetime of experience manipulating powerful men in the winner take all politics of Kazakhstan, where losing meant exile, prison, and some-times—death. He was confident he would have no trouble bending her to his will, and decided to strike while she was fatigued and anxious about her daughter.

"I'm afraid that Avalon is not well. Kazakhstan is not United States, you know. We have pathogens circulating that Avalon has never been exposed to. She seems to have picked up virus of some sort." He paused to watch Michelle's reaction, and was satisfied to see her eyes widen, her face grow pale. "So far, she seems almost normal, but this looks like maybe retrovirus, unfortunately. We are sequencing its RNA to determine whether it is new one and will know results in few days. Then perhaps we will be able to prepare vaccine to help her body conquer it."

Michelle's mind reeled quickly through the possibilities. Was he telling her the truth, that this was just a random thing? Was he bluffing for some reason? Or had he infected her on purpose? "How was she exposed?" Michelle asked.

"Well, of course one never knows. She and one of our boys decided to have adventure outside the school grounds. You know how children are, always curious. Perhaps that's when she picked it up. They are back now,"

Michelle digested this news. She knew from the telephone call that Avalon had escaped but did not know she was back at the school.

"Dr. Akenov, I want to get her home as soon as possible for treatment of the infection. We have excellent medical care at the Portland health center." For the moment Michelle decided to say nothing about Avalon's twin.

Akenov regarded Michelle, then decided it was time to break her.

"I'm afraid that will not be possible. Besides, the virus may be very infectious and you do not want to infect others."

Michelle heard the steel in his voice. "Dr. Akenov, what is it you want from Avalon and from me?"

"Professor Murphy, I am inviting you to stay here and collaborate on research project that I'm sure will interest you."

"I'm flattered," Michelle lied, but, "I'm afraid that will not be possible. I have teaching duties at home and a full research program at TransTek."

"Professor Murphy, let's dispense with formalities. May I call you Michelle? And please call me Arman. Michelle, you are intelligent woman, and I would guess you have been doing your homework, is that correct?"

When Michelle did not reply, Akenov continued. "You learned during your last visit that we have had great success constructing genetically modified horses. Perhaps you also know that this is painfully slow process, with many failures for each success, so I have also perfected method for cloning best outcomes."

He paused to gauge Michelle's reaction to the news of successful cloning, which was a state secret. When she simply nodded, Akenov smiled, understanding that she had in fact done her homework.

"What you perhaps do not know is that I direct fertility clinic here in Almaty, and I believe my duty as Kazakh citizen to extend what we learned from research on horses to Kazakh people."

Michelle's lack of surprise told Akenov all he needed to know, and his smile was real. "You have guessed what my project is. Your daughter, Avalon, is Kazakh, and she has unusual qualities that should be shared with her people. I invite you—and Avalon, of course—to help me with unique opportunity."

"No." Immediately she knew she had made a mistake. She had stepped out of character and out of his game.

Akenov continued to smile, but it had lost its warmth. "I'm sorry, I must insist. If you refuse, I doubt that Avalon will survive illness. And if she does, I do not think I could help solve these problems with the adoption papers." He paused, then added, "And as I am sure you know now, Avalon has a twin with whom she is very happy. She will not leave her twin, and of course, our officials would not want to separate two such important citizens of our country."

Michelle stared at Akenov, her mind racing. She managed to keep her face impassive, something learned from the poker she had played with her husband Hank before and during their marriage—not for money but for lovers' favors. He had also paid his way through graduate school with winnings from the Nevada casinos he visited every few months, and wanted his bride to know the thrill of using her intelligence to gauge the odds while betting against other players in Texas Hold 'Em. Michelle never reached his level of expertise, but did learn how to keep her face unreadable. She prepared herself for her next play.

"Well, Arman, what do you have in mind?"

Akenov's smile became genuine again. "I'm delighted you understand situation. It will be pleasure, Michelle, to work with such distinguished colleague."

Michelle made her face smile, but the smile did not reach her eyes.

Arman stood. "I believe you have gifts for me."

Michelle opened her purse, took out the bubble-wrapped vials and carefully set them on the table. "These should be stored at 4 degrees, but not frozen."

Akenov picked up the small packages and gestured toward a door at the far end of the room. "Then take them to lab immediately. Meanwhile, perhaps you can explain your gifts."

Akenov led the way and held the door open for Michelle. Upon entering, she was astonished to see a complete laboratory, perhaps 2000 square feet of space filled with equipment and instruments almost identical to hers at TransTek.

"Very impressive, Arman."

"You like it? This will be yours. I hardly use it."

Akenov led the way to a glove box where he inserted the bubble wrapped vials through a side port, then closed it. "I suppose you will want to unwrap these before cooling them, and it's safest here in case of accident. So. What are they?"

Michelle did not answer, but sat down at the glove box and inserted her hands into the gloves. She quickly unwrapped the vials and set them upright into a rack. Each was half filled with a clear liquid, and labeled either O or N.

"As I'm sure you know, I'm a consultant for TransTek. In my collaborative work with the company we have developed a viral vector that can transmit active genes into living cells. Furthermore, we have developed technology that allows us to target specific tissues in the human body. Each tissue has a set of receptor molecules expressed on the surfaces of its cells, and we have designed ligands on viral vectors that attach the virus only to those receptors. The cell takes up the virus after it attaches, and the virus releases its DNA which is then incorporated into the genome of that cell.

Michelle looked up at Akenov, who nodded.

"Twelve years ago I adopted an infant daughter from your country. She is becoming an extraordinary young woman, perhaps even unique. I believe you understand how unique she is, and that is why you detained her."

"I did not detain her, Michelle. Our government did. And now I can tell you that when you came to adopt her, I made that possible. In fact, while you chose Avalon, I chose you."

Michelle wanted to accuse him of Hank's death but she restrained herself. Inside her a second Michelle was raging murderously.

Akenov nodded again. He pulled over a lab stool and sat down next to her. "Michelle, since we will be working together, I will tell you truth. All my life, I have been surrounded by simple-minded people. They are easy to manipulate, but they are very boring. Until I met your daughter, there has never been another human being who could surprise me with original thought. She is treasure beyond compare. And she is Kazakh."

Michelle was surprised that Akenov would reveal personal information, but also by the clarity of his language. Apparently he could use good English when he needed to. Her response was equally clear. "Avalon is Kazakh by birth and heritage, yes. By choice, she is an American citizen." She wanted facts on the table, not an argument.

Akenov was blunt. "She is still Kazakh citizen by our choice, and she is 13 years old. So it is our choice that she stays. At least for now. How nice it will be for her and her sister to be together."

A silence settled between them.

Michelle decided to reveal more of what she knew. "When I came here to give my talk, you told us about the Kazakh horses you were breeding, and how proud you were of their successes in racing. I do understand your pride, because it's human nature. But you revealed that some of your horses are in fact clones."

She glanced up at Akenov, who said nothing but continued to smile.

Michelle thought it was time to play her wild card. "Now, you want to clone Avalon. Correct?"

Akenov laughed. "Of course. Both of them. I see that you are intelligent woman, perhaps even more so than I knew."

Michelle paused, startled by his frank answer.

"Do you have her cells in culture?"

"Certainly, fibroblasts from a skin sample. They are growing beautifully. Would you like to see them?"

"Not just yet. How did you get a skin sample?"

"Avalon needed a pertussis vaccination. I recovered a tissue sample from the needle."

Michelle considered the implications of Akenov's statement.

He broke into her thoughts. "Michelle, this is why we must work together. I have no one with your technical expertise, and I think you would take great pleasure in seeing a dozen or more Avalons growing up here at Good Shepherd."

Michelle picked up the two vials, one in each gloved hand. "Aren't you missing something obvious?" She didn't wait for an answer. "All your Avalons will be females, and cloning will take years. That's no way to produce a new race of Kazakhs."

Akenov was no longer smiling. No one had ever mocked him like this, except perhaps for Bulat when he was Avalon's age, and Bulat died a painful death.

Michelle continued to stare at the two vials she held. "Arman, these vials are not a gift. They are to be traded, and the trade is simple. You will have the vials, and Avalon, her sister, and I will go home."

"I doubt they could be so valuable that I would accept such trade. Do you really think I am that stupid?"

"Not stupid. Smart. Very smart."

"What is in the vials?"

"Two viral vectors containing Avalon's FOXP5 gene. The one labeled O will produce both male and female children, as many as you want, and the first infants will be born within a year. If you accept the trade, you will have no further need for either me or Avalon or her sister. You can have all the brilliant Kazakhs you want. Your country needs them. The world needs them. What difference if they are Kazakhs, Americans or Icelanders?"

"What do you Americans call that?" he mused. "Yes. Multiculturalism. Isn't that right?" He didn't wait for an answer. "You certainly know that thousand years ago largest empire in world was ruled by my ancestors here in Central Asia. I am not multi-culturalist." A few seconds passed while Akenov considered Michelle's offer. Then he asked, "How does your vector work?"

Michelle replaced the vials in the rack, took her hands out of the glove box and turned toward Akenov. "It's proprietary, of course. TransTek owns the IP, but I can tell you this much. We use a rhinovirus with the capsid altered to have two ligands, one for the usual upper respiratory tissues, and the second for human ova. If a fertile human female is exposed to the virus, she comes down with a cold, but some of the virions circulate to the ovaries where they bind to virtually all 300,000 ova. They deliver their DNA, which is incorporated into the haploid genome. Any egg that is fertilized carries with it the FOXP5 gene and expresses it as a dominant allele during embryonic development."

"And how much do you know about this FOXP5 gene?"

"I don't know how it acts or exactly what it does, but most of the FOX genes code for transcription factors that activate the synthesis of certain proteins. If you remember the lecture I gave, I pointed out that mammalian intelligence could be mathematically calculated as the product of two log functions. One was cell number, the other the number of synaptic junctions. My guess is that FOXP5 increases the number of synaptic junctions, and this leads to an exponential increase in intelligence."

Akenov was staring at the vials in the glove box. "If I accepted your offer, I would be buying. . . yes, I think English call it pig in poke. How do I know it will work?"

"For one thing, the O vector has already been tested."

"How?"

"I tested it on myself. I came down with a cold, and there was a distinctive pain in my abdomen a few days later when my ovaries reacted to the viral infection."

"That is no guarantee that your ova have been transfected with FOXP5."

"I suppose you must trust the reputation of TransTek, and my own expert knowledge of viral vectors."

Akenov looked down at the vials again. "What is the one marked N?"

Michelle smiled. "I guessed that you would not accept my assurance that the O vector would work, so I also prepared a second vector with a different targeted tissue."

"What is the target?"

"The N stands for neurons. The N vector will deliver FOXP5 to neurons in the cerebral cortex of adults."

"And?"

"Within 24 hours the neurons will begin to produce increased numbers of synaptic junctions."

"What do you expect to happen?"

"More junctions, more brain capacity. Put it this way—more computing power. Intelligence will markedly increase."

Akenov looked directly at Michelle, gauging whether she was telling the truth. Michelle maintained eye contact, keeping her poker face.

Finally Akenov nodded. "I accept your offer, but on one condition."

"And that is?"

"You will test the N vector on yourself, before me, right now."

Michelle shook her head. "Too dangerous. Any new vector must be tested first in animal models. We use chimps at TransTek."

"Then you must stay and help me test the O vector. We can use patients at my fertility clinic."

Michelle willed no muscle to move in her face, and she hoped her eyes did not betray her astonishment that Akenov would immediately go to human testing, bypassing all of the usual animal trials.

"How long will that be?" she asked.

"At least year, I am sure."

"What about Avalon?"

"What about her?"

Michelle'd heart skipped a beat. Akenov might have made a mistake.

"Her infection?"

Akenov looked away. "Oh, yes, we will immediately begin treatment."

Michelle decided that Akenov had been bluffing, that Avalon was not really sick. She decided to play her ace.

"We can't wait that long. If you promise to begin treating Avalon, I'll test the N vector. I will take that risk, because I think the risk is small."

Michelle watched Akenov's face as he considered this. A minute passed. Finally he nodded. "This a mother will do," he said as if reminded of a basic scientific principle.

Michelle said to herself, *You have no idea what a mother will do.* She put her hands back in the glove box and picked up the vial marked N, deftly breaking the top off to open it. "I'll need a cotton swab and a sterile vial to store the vector."

"Swabs and vials are in box, on shelf to right."

Michelle found a swab and touched it to the surface of the clear liquid. She set the swab upright in the rack, then poured the rest of the liquid into a vial and screwed on the cap. "Arman, please put on some gloves and bring the rack out through the port. Give me the swab, but the vials can go into cold storage now."

She watched as Akenov followed her directions. When he returned from the refrigerator, she patted the lab stool, indicating that he should sit. She turned to face him, inserted the swab into the back of her throat and moved the cotton back and forth for a few seconds. Then she asked, "What do you do with biohazards?"

Akenov pointed to the next aisle where Michelle saw a white container with the familiar biohazard symbol on it. She walked over, deposited the swab, and said, "That's all there is to it. Now I want to see my daughter."

Chapter 34

Akenov rose from his stool and opened the door that led from the lab to his office. Michelle was surprised to see Asel waiting there.

"Asel, please take Professor Murphy to dormitory where Avalon is staying. Classes are over for today, so I imagine she will be there."

Asel nodded and smiled broadly at Michelle. "This way please."

Michelle was certain this had all be scripted before her arrival. Neither woman spoke as Asel led Michelle down the now-familiar stairs and through the steel door that led out into the playground.

Asel asked, "Is everything alright?"

Michelle shook her head. "I can't talk about it yet. It's complicated."

Asel sighed. "Well, at least you and Avalon will be together again. And you will meet her twin, Aidana."

Michelle stopped walking, partly to make a point, but also thinking she might be able to change the script of the play in which she was an actor. "Asel, you knew about her twin. You didn't tell me. You're the first person to say her name. Adaina."

Asel stopped and looked back at Michelle. Asel corrected her syllable by syllable, "Ay da NA. Yes, I did know. But please understand me. Sometimes it is better not to say too much."

Michelle relented when she sensed that Asel was talking to her as another mother. She said cheerfully, "That's probably good advice, but I'm Irish and the Irish like to talk."

As they walked across the deserted playground, Michelle had to shade her eyes against the bright sunlight. A faint chill lingered in the air, even though it was early afternoon. Michelle saw the equations painted in large white letters and asked Asel what they were.

"Some sort of test, I think, that the director made up. Avalon knew what they were.

Michelle smiled and said, "She would."

They had reached swinging glass doors that led into a two story brick building. Asel pushed them open for Michelle to enter. "This is the girls' dormitory. The boys have their own dorm on the other side of the play yard."

"How many children are here at the school?"

"We're limited by the number of dormitory rooms, so just over two hundred. Here we are, by the way. Avalon and her roommates are in 103."

Michelle, still jet-lagged, felt a little faint, and leaned against the wall as Asel tapped on the door, then opened it and looked in. She saw Avalon and Nurzhan standing at the blackboard, discussing what looked like a mathematical expression. "Avalon, your mother is here."

Michelle heard a gasp, and pushed her way past Asel into the room.

"Mom!" Avalon ran across the room and threw herself into Michelle's arms, then began to sob.

Michelle, with tears running down her cheeks, could only hug Avalon and murmur, "It's OK, I'm here." She was surprised, because Avalon had never before shown this much emotion. Through the tears she saw the other Avalon calmly smiling at them, then another pair of arms circled her waist from behind and she heard a girl's voice saying "Ma, Ma, Ma."

Avalon laughed and was finally able to catch her breath with an occasional hiccup. "Oh, Mariam, stop it. This is MY mother. Mom, this is Mariam. She's very affectionate, as you can see. She's also a musical genius. And that's Nurzhan, my other roommate over by the black white board. She loves math even more than I do. And, Mom, that's not me at the board, it's my twin sister Aidana."

Asel, who had been watching the reunion, gently unwound Mariam's arms and led her over to her bed, where the girl sat down and began to hum to herself. Aidana did not understand English, but was beaming with happiness for Avalon. In only 2 days they had grown very close as they discovered minute by minute how much alike they were, and that they were no longer alone in an existence no other girl, boy, woman or man shared.

She looked at Michelle with Avalon's eyes. In fact, Michelle was sure she saw one of Aidana's eyes, the left one, look at Avalon while the right looked at her. It was a brief fraction of a second, then both eyes held hers and somehow pulled Michelle toward her across the room as if the air between them were disappearing. Michelle stretched out her arms and grasped the tall girl's shoulders and looked in her eyes. Here was Avalon who had never had parents or lived in America. Michelle smiled warmly and drew Aidana to her. For a long moment they held each other. Aidana said, "Avalon's mother, my mother."

Avalon said, "Mom, can you believe it? Aidana and I are going to be co-stars in a school play in two days. It's kind of silly stuff but we're the heroines and Marat is the hero."

Michelle's thoughts were far afield, not on a school play, but on her role in a drama that had been going on far too long. She wanted to be off stage. She gave Aidana a final squeeze, said, "I'll look forward to it," and released her.

Michelle turned to Asel and said, "Could you take us to my room? I need to talk with Avalon, but I'm not sure I can find my apartment again."

Avalon said something to Aidana in Kazakh, and Aidana replied with in heavily accented English, "Yes. Understand." To Michelle Avalon said, "Mom, I know where it is, upstairs and down the hall from the library, right Asel?"

"That's right, but this morning I forgot to give your mother the key." Asel pulled out a jingling bunch of keys that hung from her belt by a chain.

Avalon, hearing the keys, said, "Tintinabulation of the keys, keys, keys."

Michelle smiled and almost automatically joined the game that had started before Avalon could read. "Poe—Bells," Michelle said. Avalon smiled.

Aidana chimed in, "Tintinabulation of keys, keys, keys."

Asel looked puzzled as she detached a key from the ring. She offered it to Michelle, but Avalon grabbed it, took Michelle by the hand and said, "Let's go Mom!"

As Michelle was being pulled through the door, she caught Asel's eye and managed to say "Thanks for helping. I really appreciate it."

Then Nurzhan yelled, "Fibonacci!" and pointed to the numbers on the board.

In Kazakh, Avalon said over her shoulder, "Later Nurzhan. I'll be back. I need to talk to my mom."

As they climbed the stairs to the apartment on the second floor, Avalon took Michelle's arm to stop her, then whispered, "Mom, we can't talk in your apartment. He might be listening."

Michelle thought for a moment, then nodded and signaled by pointing to Avalon, then to herself and making walking movements with her fingers, one set following the other, meaning: *Just follow my lead.*

In the apartment, Michelle went to the kitchen and opened a large bottle of orange soda from the refrigerator. She poured two glasses and carried them back into the living area where she motioned Avalon to sit in a chair opposite the sofa she sat on.

"Sweetheart, there's something I need to tell you." She took a sip of her soda, sighed, and said, "We're going to be here awhile longer."

Avalon was genuinely surprised. "Why? I thought you were here to take me home."

Michelle winked at Avalon to let her know she wasn't serious. "Well, I agreed to help Dr. Akenov with his research. Did you know that this place is more than just a school?"

Avalon was now in the game. She pretended. "No. What do you mean?"

"Dr. Akenov is very smart, and this school is his experiment. He was hoping to find super-intelligent Kazakh kids like you and bring them all together where they can grow up and maybe help each other develop mentally to the fullest extent. He's hoping that the smart genes would become concentrated and passed along to their children when the kids grow up, something like he has done with fast genes in race horses."

"I don't think it's working, mom. I only met one kid who comes close, and that's Marat. Some of the others, like Nurzhan, are very smart, but in only one area like mathematics."

"Yes, Dr. Akenov knows that, which is the reason he is so interested in you."

"Why me?"

"Because you are smart in all areas. And I suppose I know why."

Avalon said what the poet Frost wrote, "We dance round in a ring and suppose, But the Secret sits in the middle and knows."

"Not any longer," Michelle said. "I never told you this, but when you were two years old I analyzed your genome. You have a new gene that I'm calling FOXP5, and it makes a protein called. . ." Michelle paused, then smiled and said, "Avalin." Michelle nodded her head and opened her eyes wide to say this was true.

Avalon couldn't help laughing. "Really? Have you told anyone?"

"No, but I did publish the sequence, because at first I thought it had something to do with autism. It doesn't, but Dr. Akenov happened to see my paper and realized that FOXP5 might be important. We both think it could be responsible for your intelligence."

"Is that why I'm here?"

"Yes." Michelle winked again. "Dr. Akenov believes that unique genes, like minerals, oil and other natural resources belong to the government of Kazakhstan. I can't win a legal argument about this in a Kazakh court, so we made a deal. I will collaborate in his study of FOXP5. Having two of you with the same gene but different lives is a real bonus. When we understand the composition of FOXP5 and how it works, you and I can leave."

Now Avalon winked at Michelle. "And Aidana can come with us."

This caught Michelle off guard. She answered honestly. "I know you would like that very much, and we can talk to Dr. Akenov about it."

"I think I can explain it to him so he will understand," Avalon said.

Michelle didn't know where this was going, and she shook her head and motioned with her fingers that Avalon should not pursue this. "

Avalon spread her palms and bowed her head to say she understood but was not happy.

"I hope he will understand," Michelle said. "But let's not assume he'll agree with you."

Avalon thought for a moment, then asked, "What do I need to do?"

"Not much, actually. You can continue with your classes here and enjoy your new friends, and Aidana of course. Your overall intelligence has never been measured, so we will give you some tests to establish where you are on the IQ scale. We will also have you scrape some cells off the inside of your cheeks so that I can repeat the genome analysis I did ten years ago. We know that you have the new FOXP5 gene, but I never determined how many copies there are, or which chromosome has it."

When Michelle was done talking, she glanced toward the door to indicate that she had said enough. Avalon took the hint, and said, "Mom, it's such a nice day, let's go sit outside for a while."

"Good idea. Take your glass, and I'll bring the soda. I'm thirsty for some reason, maybe jet lag."

Akenov had, in fact, been listening. There were hidden microphones in every room of the school so that he could monitor student conversations for signs of unusual intelligence, as well as visiting parents in case any of them became suspicious about the school. He was content with his sense of control and secure in his knowledge. Unfortunately, his students bored him. He was a disappointed father. He deserved better.

He went to the window that looked out onto the play area, and a few seconds later he saw Avalon and Michelle walk to a bench where they could sit in the sunshine. He had misgivings about what he had heard. It sounded a bit contrived, but in fact Michelle had told Avalon nothing but the truth. He would accept it for now. He might know more tomorrow when the initial effects of the N vector in Michelle could be apparent.

Out on the playground, Michelle and Avalon huddled together, speaking in barely audible voices.

"Avalon, I can't tell you the reason yet, but tomorrow when I talk to Akenov I need to pretend that I'm as smart as you are."

"Mom, you don't need to pretend. You're very smart."

"Thank you, but experience is not the same as intelligence. I'm 30 years older than you. You know as well as I do that you understand things at a much deeper level than any other 13 year old, and most grownups as well."

Avalon considered that for a few seconds, then nodded and said, "OK, what do I do?"

"I need to guide my conversation with Akenov so that he asks only questions I can answer. Don't look up, but a good place to start is a kind of test he set up with those equations on the wall. Have you talked to him about what they are?"

"Not really. I did talk to Asel and Marat."

"Well, I need you to tell me what you know about them. I'll memorize as much as I can, and you can test me."

For the rest of the afternoon, Avalon explained the equations to Michelle, then took her to the cafeteria for dinner around 6 p.m. When they pushed through the glass doors and walked to the service counter, each table of students stopped their noisy chatter and followed Michelle with their eyes. It was unusual to have a grown up in the cafeteria, except when the occasional parent visited, and they were puzzled that an obvious Kazakh student would have a very white skinned, round eyed woman tagging along. At the service counter, Avalon explained the exotic dishes to her mother, helping her to choose ones she might like.

With trays full, Avalon led the way to the far end of the room near the windows. She looked around for Marat, but he had not yet come in. "Genghis Khan has not come in today," she said.

"Genghis Khan?" Michelle asked.

"In the play—night after tomorrow," Avalon reminded her. "How do you like the title: 'We Are All Genghis Khan'?"

"That leaves me out," Michelle said. "Is there a role for Queen Maeve?"

"The queen sits in the audience," Avalon giggled. "They also serve who only sit and wait."

Michelle smiled and said quietly, almost sadly, "Milton—who only stand and wait."

Avalon added, "God doth not need either man's work or his own gifts: who best bear his mild yoke, they serve him best."

Michelle decided that she and Avalon should expand on her original research field, astrobiology, in which she had significant expertise. They discussed how life can begin on the surface of a habitable planet, and the possibility that thousands of planets in the galaxy harbored something that could be called life. By the time Avalon got to the math of the Drake equation and the likelihood of intelligent life elsewhere, Michelle began to lose the thread and called a halt. "Avalon, I've got to be fresh in the morning, so I'd better get some sleep."

"OK mom, but it's dark out now. Do you need help finding your room?"

"Thanks, A-stór, but I remember how to get there. Asel will come fetch me in the morning so you don't need to worry. Go get some sleep yourself."

"Wait. Astoor? I never heard that before. What does it mean?"

Michelle laughed. "I don't know how that popped out. In Irish, it means my treasure. It's what my mother used to call me."

Mother and daughter stood and hugged. While they were in the embrace, Michelle said softly, "Before you go, there's something best done in the dark. Come outside with me."

A few students lingered in the cafeteria and they paid little attention as Michelle and Avalon pushed through the doors. Michelle said softly, "Do you know who the man in the black leather jacket was standing near the kitchen door? Don't look but another one is standing just inside the door to the gatekeeper's booth."

Avalon didn't turn her head, but her left eye looked briefly toward the gate. "They work for Akenov. One of them was in the car that came to the church when we were there."

"I think they're here to make sure we stay put," Michelle said.

"Then they should be happy," Avalon replied. "I wasn't planning on going anywhere."

They sat again on the same bench they used that afternoon. Even though the playground was empty and quiet, Michelle spoke almost in a whisper. "Avalon, do you remember seeing a gold horse in Akenov's office?"

Avalon nodded. "It looked like the one dad showed you in that book."

"I saw it today for the first time, and you're right. It's the same one, or a copy. Is it just a coincidence that Akenov has it?"

"He works with horses, so it makes sense that he would collect one. But did you notice that this one has the horse running in the opposite direction? Maybe it's the original and someone imitated it, but made it different so that no one would realize it was a copy. Only someone like Dad would know it was a fake."

"Well, let's keep thinking about it. But now I need to talk to you about something very important."

Michelle went on tell Avalon about the SV40 virus, that the CIA knew Akenov had worked on it and that was why she had been invited to the conference. On this trip, at Lipkovich's insistence, she had brought with her a simple skin patch test for the SV40 virus.

"I want to test you and I want you to test Aidana."

Avalon nodded. "Sure. Whatever."

The test was an improved version of the standard allergy test. A scratch on the upper arm, a small bandage smeared with antigen, and if the person had been vaccinated, the skin would begin to turn pink in an hour, while unvaccinated skin would show no reaction.

"I'll give you another scratcher and a bandage with antigen on it and you do the test on Aidana if she'll let you."

With the bandage in place where it would be under Avalon's blouse, Michelle kissed Avalon's hair. "Good night darling. I can't tell you how good it is to be here with you. I think we're going to be all right."

At the dorm they said goodnight with another hug.

Chapter 35

Next morning a gentle tapping at Michelle's door penetrated her deep sleep. Not sure whether the noise was real or in a dream, she automatically called out, "Just a minute!"

She sat up in bed, stretching carefully to avoid painful cramps. Sleep had been an escape, and only while she stretched did the memory of where she was and what she was doing come back. *The prisoner-spy is ready* she told herself. She plodded to the door and opened it a crack. Asel stood a step back from the threshold, holding a tray. In her white technician's uniform she stood like a marble statue with a brown face and black hair.

"Good morning. I think you might like to have a small breakfast here, rather than the cafeteria. It's quite noisy in the morning."

"Oh, Asel, that's very thoughtful of you. Come in. What time is it anyway?"

Asel followed Michelle to the kitchen and set the tray down on the small table, careful not to spill tea from the dark blue porcelain tea pot with gold Kazakh designs, the same one Michelle had seen in Akenov's office. She consulted her watch. "Almost 11. You had a long sleep."

"Oh, my! I must have been really tired. That tea looks good, and where did you find croissants?"

Asel smiled. "Not in the cafeteria. There is a little bakery near my home, and they have learned to make croissants in the French style." Nodding down at a very large apple sliced into crescents, she said, "The apple is our own Aport, a very old apple from the days when Kazakhs called this city Alma Ata, ancestor of apple."

"Yes, I know the story," Michelle said. "Stay and help me eat all of this—it's too much."

"I wish I could, but Dr. Akenov is waiting for me. He asked me to invite you to his office at noon. Is that alright?"

Michelle thought, *Invitation, my eye.* She said, "Of course. That's plenty of time for breakfast and a quick shower. Will you come to fetch me when it's time?"

Asel nodded. "Yes, of course. Please enjoy your breakfast now."

"Before you go, Asel, can you tell me something?"

"I'll try."

Michelle walked to the kitchen window from which she could see the entry portal. "The man in the black leather coat at the entry—who is he?"

Asel did not come to the window and said nothing for a few seconds. "I don't know them personally." She paused again. Almost secretively she said, "I don't like them. They are Dr. Akenov's guards."

"And mine too?" Michelle pressed.

"I think so," Asel said, "But you know I must be careful."

"I understand," Michelle said. "See you later."

After Asel closed the door behind her, Michelle nibbled a croissant, found it a little rubbery but acceptable, and took a sip of tea. She considered what to do in her meeting with Akenov, feeling something like an actress facing an audience for the first time. She had always been entirely honest with friends and colleagues, so putting on an act was as strange as submitting someone else's resume and calling it your own. She was learning what had to be done.

After a second croissant, Michelle found a pepper shaker and sprinkled a pinch into a handkerchief that she put in her purse. Then she stood for several minutes in a hot shower willing herself to relax. After drying off, she dressed in a conservative gray pantsuit, the same one she had worn for her lecture just a month ago. She brushed her hair, but did not put on makeup. Finally she left the door ajar and settled down to study the notes she had made yesterday while talking to Avalon.

Half an hour later Michelle heard Asel's footsteps coming down the hall. She went into the bathroom, and when she heard a knock on the door she called out, "Come in! The door is open." Bracing herself against the sting, she put a tiny drop of shampoo at the edge of both eyes and used a towel to spread it around. She wetted the towel and wiped her eyes, which had become red and overflowing with tears.

Asel was surprised when Michelle emerged from the bathroom. "What happened to you? Are you crying?"

"No, no, I think I caught a cold on the plane and it's just starting to set in."

"Shall I tell Dr. Akenov that you're sick?"

"I probably look terrible, but I don't feel sick, at least not yet."

Asel was skeptical, but shrugged and said, "Well, alright, but be sure to tell Dr. Akenov if you don't feel up to it. We can go there now."

When Asel knocked on the door of Akenov's office, Michelle took the handkerchief out of her purse and was holding it to her nose when the door opened. She breathed in the pepper, and stood there for a moment until the sneeze came.

Akenov was standing behind his desk, watching. "Are you all right?"

Michelle shook her head, then glanced sideways at Asel. Akenov took the hint. "Asel, I need to speak with Professor Murphy in private. You are needed over at clinic."

After Asel left the room, Akenov walked to the couch by the window, saying, "Asel is a very useful assistant." He beckoned Michelle to join him. She sat in a comfortable chair with the table between them, wiping her eyes and sneezing again into the handkerchief. Akenov asked, "Is this effect of vector?"

"Of course. The vector is a rhinovirus."

"Oh yes, I see," Akenov said. "I know these cold viruses. Highly infective." He stared at Michelle, noting her reddened eyes and watching her blow her nose. "Do you feel any other effects?"

Michelle put her handkerchief away and looked directly at him. "Yes. It's working." And she began to feel confident that it—her plan—was working.

"How do you know?"

Michelle shook her head. "How do you explain color vision to someone who is color blind? I feel like Dorothy in the Wizard of Oz, after the tornado landed her in Munchkinland, when she stepped out of her black and white house into a technicolor world. Have you seen the movie?"

"Years ago. I prefer Kazakh films. What else do you feel?"

"It's really quite extraordinary. Yesterday I was living in a three-dimensional world. This morning I woke up to four dimensions. Einstein must have had the same vision when he added time as a fourth dimension."

"Do you think that Einstein had FOXP5?"

"I doubt it. He made a few great discoveries in a narrow field—a genius in physics and math, but rather ordinary in other sciences, politics and social life. I think he was at the far end of the bell curve that describes variations in normal human intelligence. He must have been a lonely child."

Akenov nodded. "No friends," he said. "I know about that." He adjusted his position on the sofa. "At my end of bell curve very few friends. But Avalon—she could be my friend. And Aidana too. Yes, they could, you know that." He stared intensely at Michelle. "Maybe you too." He paused. "Am I right?"

Michelle looked out the window, ignoring his direct question, summoning the will power not to jab back. Two pigeons flew past, and she watched them land in the upper branches of a tree. "Everything is surprising me. Instead of seeing just the presence of an object, and its surface, I see how the object relates to the rest of the world and how it changes over time. This feeling of understanding starts with the cosmos and goes right down to the subatomic level. For instance, you have put twenty or so equations on the walls surrounding the playground. Some of these are mathematical, such as the one that goes 0, 1, 1, 2, 3, 5, 8, 11, 21, 34 and so on. I don't know what it's called, but each number is the sum of the two numbers preceding it. By the way, the 11 should be a 13 so you should correct that. The thing is, when I saw those numbers, my mind immediately brought up images of ferns, pineapples, pine cones and artichokes. Without realizing it, in the past I had seen organized structures in plants, and now I know that they are described by those numbers."

"Yes. Numbers discovered nearly thousand years ago. Fibonacci series. But this is known. You are seeing what is known."

This was the opening Michelle had been waiting for. "You are skeptical, but here is another thing I now understand. Yesterday Avalon and I went to the cafeteria for dinner, and I noticed multiple sets of twins there and even triplets, far too many to be accounted for by chance. I concluded that the children are not admitted at random from the Kazakh population, but that most of them are the result of *in vitro* fertilization at your clinic."

Akenov was startled, and took a minute to respond. "Yes, as a service to the parents who are my clients, I invite them to send their children to Good Shepherd when they are old enough."

"In vitro fertilization is expensive, and only a few Kazakh families could afford it, most likely business leaders and powerful people in government. Is that right?"

Akenov became very cautious. He tried to think ahead but could not see her goal. "Yes, I'm proud to say that the vice president and head of security are both clients. They have been very helpful in contributing to our endowment at Good Shepherd."

"I wonder what they would say if they knew you were the biological father of their children?"

Akenov leaned forward to the edge of the couch and looked at her very hard, but the look was a dispassionate examination from head to toe. For the first time in this game, Michelle sensed imminent danger. To him she was not a human being. She was a thing to be used or dealt with. She took out a handkerchief from her bag and dabbed at her eyes. The remaining pepper made both her eyes and her nose run. She blew her nose. "Excuse me. The price the virus extracts."

Akenov also felt danger but long ago he had trained himself to tolerate surprise and sudden disorder. He triggered his self-control by silently repeating a name—Bulat, Bulat, Bulat. Bulat was dead. Arman was alive. That was as it should be. The right order. What he could not tolerate was someone much smarter than him, especially not in his own field.

"Michelle," he said in a voice he might use with a student or a lackey. "You should not come to Kazakhstan to cause trouble and speak of things you do not know. You are our guest. My guest. Do you understand?"

Michelle no longer knew what the consequences of continuing would be, but she knew she could not give up control so easily. "The pattern is obvious, Arman. I'm a geneticist, and I see your dominant traits scattered throughout the children here. Even Marat. Does Asel know you are Marat's father?"

Akenov said nothing for a long time. He stared intently at Michelle, trying to bore into her thoughts, her goal. Michelle looked back as calmly as she could. She saw nothing in his eyes. He turned away like a man who has been looking for something inside a cabinet without finding it. She believed she was in control, but she did not know for how long, and she did not know what this small man with very big ambitions and visions might be capable of. She decided to play her trump card and picked up the medallion.

"Dr. Akenov, when I saw this, I immediately understood that you caused my husband's death. He exposed one of your counterfeit antiquities you were selling, an imitation of the gold medallion that I am holding. You took revenge by infecting him with a viral vector delivered by a modified SV40 virus, probably as a weaponized powder that was contained in a package he would open."

This brought Akenov to his feet. His eyes were blazing, his confidence shattered. Michelle looked around for anything that could serve her as a weapon if it came to that. She calculated how fast she could reach and open the door.

Akenov saw this and said, "You want to leave. Go."

"Gladly. You have the vectors. Can I take Avalon and Aidana home now?"

"You will leave." It was no longer a request but a command.

Michelle knew she had won this round. She placed the medallion back on the table and stood up. "I want Avalon and she wants her sister. You don't need them."

Akenov turned his back to her and said again, "Leave now." He added, "And do not underestimate me. I am not alone."

Michelle quietly left the office, leaving the door open and unsure if he could hear her depart. Asel had disappeared, so Michelle went down the stairs alone and pushed through the door to the playground. Despite her calm performance, her heart was beating madly, and her palms were slippery with sweat. She leaned back against the door and took several deep breaths, worked her shoulders and neck muscles loose, then walked across the playground toward the dormitory. As she passed the gate and guard post she waved flirtatiously at one of the men in a black jacket. She thought she saw a slight and puzzled smile stretch his lips.

Chapter 36

After tossing and turning most of the night on the hard mattress of the bed in her apartment, Michelle finally slept for an hour in the early morning. She woke up when she heard the door open, and could see Avalon peeking in from the hall, Aidana behind her.

"It's OK sweetie. I'm awake."

Michelle was still groggy from the stress of her confrontation with Akenov, feeling like someone who has set a mouse trap but not particularly looking forward to the result.

"Hi mom. Want to come have breakfast with us?"

"Sure. Just give me a couple minutes to wake up." She shook her head and smiled at Avalon's good mood and nonchalance even while they were both prisoners of a dangerous man. Was it her intelligence or her youth?

Aidana was sitting with Avalon at the small table in the kitchenette, speaking softly in Kazakh while Michelle showered and got dressed. Fifteen minutes later she was ready, and they walked downstairs into the play yard. No sun shone on the well used ground today. A chilly mountain fog had descended, hinting at the cold fall weather soon to come. Aidana led them into the dining hall where a scattering of sleepy students were having breakfast. There was no line, so they quickly supplied their trays with kasha and scrambled eggs. Avalon waved at Marat, who was sitting alone at an empty table, and she and Aidana walked over to join him. Michelle was following with her tray when she glimpsed someone hurrying across the play yard toward the dining hall. Asel burst through the doors so suddenly they banged against the wall. The usual breakfast clatter faded away, all faces turned toward the sudden noise. In the silence, Michelle heard Avalon quote a Belloc poem: "A trick that everyone abhors/ in little girls is slamming doors". But then Avalon turned and saw Asel's ashen face. She realized instantly that frivolity was out of place and said, "Sorry."

Michelle put her tray down when she realized that Asel was holding the door open, making it obvious that she should come. Michelle looked down at Avalon, Aidana, and Marat and whispered, "You three stay here." Marat shook his head and started to get up but then sat down when he felt Michelle's hand press on his shoulder.

Michelle walked quickly to the door, and Asel led the way across the playground, saying nothing. She used her key to open the door to the office complex. Michelle took her arm to stop her. "Asel, what's wrong?"

Asel did not turn or speak. She pulled free and ran up the stairs two at a time. When they reached the long hall, Michelle saw the open door to Akenov's office and a man she didn't know standing in the doorway.

Asel stopped, panting a little. "Michelle, this is Nurlan, Dr. Akenov's secretary."

The young man with the sad Russian eyes looked at Michelle and nodded.

Michelle nodded back, then turned to Asel. "What is this about? Where is Dr. Akenov?"

"He's here. Nurlan, tell her."

Michelle shook her head in confusion. "You speak English too?"

"Yes, of course. I am Dr. Akenov's translator. I translated your papers for him."

"What's happened?"

Nurlan stood aside and Michelle stepped into the room facing the desk.

Akenov was not at his desk. She looked over to the furniture suite and saw his distinctive black hair over the back of his favorite leather upholstered couch. He did not move. Michelle stepped cautiously to where she could see his face and small figure. His eyes were open but unfocused. From his open mouth she saw a thin line of drool led down his chin to a damp spot on his jacket. She was not a physician, but she knew enough about neurology to run through a series of simple tests. Akenov's pulse was steady. He did not move as her fingers touched his wrist. He breathed very slowly. His pupils were of equal size and not dilated. She used a fingernail to scratch the back of his hand and saw no response.

Michelle looked up at Nurlan and Asel, who were standing beside the couch.

Asel spoke. "I didn't call Skoraya Pomosh, quick help—you call it what, ambulance or emergency service—because he had always told us if anyone is contaminated with anything, learn first what it is, because our doctors make too many mistakes. I thought he had a stroke, but he did say he was testing something. Maybe it was on himself."

Michelle was trying to decide what to do next, then Nurlan spoke.

"Last night, Dr. Akenov called me, saying it was urgent. I thought maybe he had to send something to the west where the business day was still going on. But he dictated a note addressed to you, then told me to leave and return in the morning. When I arrived, I found him like this."

Asel said, "I will call for a doctor, an ambulance."

As Asel moved toward the phone, Michelle said, "It's too late. If he's dying, let him die in peace."

Michelle looked at Nurlan, and to her surprise, he nodded his agreement. Michelle thought quickly. Could Nurlan be on her side? What about Asel? What would convince her? Then she knew the answer.

"Asel, for the children—inside these walls they are safe, at least for a while. Outside probably Akenov's men are waiting, his Kazakh Shepherds. We don't know what they would do if they knew he is dying. They are not our friends. They are not your friends. Do you understand?"

Asel covered her face with her hands.

Nurlan said, "She is right, Asel."

Asel, still covering her face with her hands, said, "All this time I didn't want to know. But I knew. Didn't I?"

Neither Michelle nor Nurlan answered. Then Nurlan shrugged and said to Michelle, "Dr. Akenov left a message for you on his computer, something I translated because he wanted the English to be correct."

"Can you print it please?"

"There is no printer here. Everything he wanted to print he sent to my computer, but that requires his passwords."

"But the computer is on," Michelle said. "We can read it on the screen. Put it on a flash drive."

"We can read it but any download required his password, which I don't know."

Nurlan tapped the return key and the screen lit up with Akenov's letter. Michelle looked again at the small figure of Akenov, now curled in the fetal position but still breathing. She turned to the screen and began to read.

"*To Michelle Murphy.*

I am dying—as you well know. This letter is my good-bye and perhaps your death sentence. You have ten minutes to read it, then it is wiped from memory. When you finish reading, the score will be two for me, one for you.

Michelle stopped reading. Two for Akenov? She desperately tried to highlight the entire letter and to paste it into a new file, but the copy function would not work. She continued reading.

Do you know the penalty in Kazakhstan for murder? What you have done I have described in precise detail, and those details I have sent to responsible people in my organization. And, of course, without either American parent, Avalon's only home will be in Kazakhstan. We could say that what Akenov gives, Akenov can take away.

That was all the proof she needed, but the proof would soon disappear. She tried the copy function again, without success. She almost shouted, "Cell phone! Who has a cell phone?" Nurlan pulled a phone from his pocket.

"Camera!" Michelle commanded. Nurlan handed it to her with the camera function ready and she began to snap pictures of the screen as she read through the text.

Your N vector worked, but it is working too well. My ability to think, my perception, these change even as I speak. It is a remarkable sensation. The brain is growing, intelligence is evolving. Usually the simple growth of the human brain takes years, and we do not feel the difference from one day or one month to the next, but I feel it happening in a matter of minutes, an hour at most. I can feel it happening, and now I understand clearly what you are doing—what you have done, I am sure. You tempted me with the promise of superior intelligence. You knew that I could not resist. You hid your intelligence behind an apparent dullness.

That phrase, "apparent dullness" jabbed a finger in Michelle's ego. She set aside the insult and read on.

"That doesn't matter now. You suspected, and I now know that the rapid growth of synaptic connections occurs too fast for an adult brain, that the synapses have insufficient time to arrange themselves on the cell bodies, so they form new, random connections. The result is a rapid onset of a condition resembling autism. Now my mind is constantly in danger of falling into a deep groove in which it can focus only on a single thought. This is how my daughter Nurzhan must feel as she becomes endlessly fascinated with a mathematical theorem."

"As your song says, 'Once I was blind, but now I see.' I understand why you were not harmed by the N vector. You had tested the same rhinovirus on yourself with the O vector, which meant you were immune to it, just like a vaccination. Very clever! The neuron targeted N vector could not produce an infection. Yesterday I did not see the evidence of your deception, your act. You could not have understood the symbols on the wall, but Avalon did, and she must have you told you what they meant. You are not a very good actress, but you had the advantage of an autistic audience who wanted to believe."

"I am slipping away, Michelle, and Nurlan is wondering why I am speaking so slowly. Such an effort to drag my mind up out of a groove! But I am strong. A few last things before I disappear. Your great mistake is that you should have realized what we could have done working together. Our new president is my friend. My name is famous and I have an organization with real power. We could have assured my future as the leader of a new Kazakhstan. Surely you understand that for the future of the world, science must also be power. And did you not know that Avalon and Aidana could have been my heirs and the first of the new human race? They could have been the most powerful women the world has ever known."

"Would you like to know who their parents are? You, or more likely Avalon, guessed correctly that I am the father of half the children here at Good Shepherd. Like Genghis Khan, my genes will live on in the Kazakh population, but not in Avalon and Aidan, because I am not their father. Fourteen years ago, one of the

nurses in my clinic came to me for help. Her 15 year old daughter had a brief romance with an older boy. When the daughter became pregnant, the boy disappeared. She didn't tell her mother until it was too late, so the girl gave birth to twin girls in my clinic. She was too young, too small a pelvis, and a Caesarean was necessary. There was great loss of blood, the girl did not survive and her mother wanted nothing to do with the infants who had killed her daughter. I arranged for them to be cared for at the orphanage, but a few months later the director called me to say there was something very strange about the twins. Their eyes could look in two different directions and they seemed very precocious, paying much more attention to their surroundings than other infants. Then an older couple in Astana adopted Aidana a few weeks before you arrived, and you adopted Avalon. I followed Aidana's progress as she grew up and guessed that she and Avalon might be unique. Then I happened to see your paper about Avalon's unusual gene. It is ironic, yes? I had the next human species in my hands without realizing it, and let them go! I had to bring Avalon here to compare with Aidana. Nature versus nurture. I'm sure you understand my curiosity. Her genome was also, as they say, a backup copy."

"Now I believe you have already given the human race what I planned to give only to the Kazakh people. We are both playing God, I think. Do you agree? That has always been human destiny, that we would become gods. I, however, was discriminating. Like Genghis Khan, I had a chosen people, chosen once by evolution and again by me. I chose a natural foundation of Kazakh genes. You chose everyone, which means you chose no one. You have not controlled the experiment, but you abandoned it to nature, to chance. I realize now that the only way for you to defeat me was to broadcast the vector as you traveled to Kazakhstan. That was the atomizer in your luggage, filled with the O vector. You have won, and I have lost. I congratulate you on your victory."

"Michelle, now you have two choices. If you choose incorrectly, you will be arrested and tried for murder. The penalty for murder is death. But the correct choice involves Avalon and Aidana. They belong to Kazakhstan—our national resource, not yours. As soon as you decided to return for her, I prepared the papers for you to give up all claim to our national treasure, which in any case you do not have legal right to. These papers are here in possession of my Good Shepherds. If you choose correctly, they will absolve you of my murder and give you safe passage. Avalon and Aidana will stay and soon be celebrated as the mothers of a new human species. Our last collaboration shall be to name this new human species—Homo omnipotens? Better, Homo divinus who shall soon have all the powers once attributed to the gods."

This letter I have stored for the historical record. I want the world of superior humans to know about the woman who created them and who murdered Arman Akenov.

Your colleague and your adversary, Arman

Michelle touched Asel's arm and pointed toward the screen, inviting her to read. Michelle walked to the window and stood watching children in the playground. Two minutes later she heard Asel say, "I couldn't finish it. The screen turned black and the computer shut down."

"Don't worry. I have it all on Nurlan's cell phone."

Asel asked, "Did you know this was going to happen? To Dr. Akenov?"

Michelle shook her head. "The vector has never been tested on humans, so there was no way to predict what would happen. All I could have said is that it would be very dangerous."

"Did you warn him?"

Michelle continued looking out the window. Somehow the answer was in the random music of the children's play, but she could not find the words. She said the simple but useless truth, "No."

"What did Dr. Akenov mean, that you broadcast the vector. Did you?"

Michelle looked across the playground toward the gate. "No. I thought about it, even brought an atomizer to spread the virus, but I could not do it. I have different reasons than Dr. Akenov." She saw at the gate the smaller of the Good Shepherds, the intent little man with the pinched face. He looked up at the window and pointed at her. He seemed satisfied. His look brought her back to the present and she turned away from the window. "Asel, we are in danger here. You can't imagine how dangerous Akenov is. Or was, I suppose. But he still has his so-called Shepherds." She turned to Nurlan. "Did he say anything else after this letter. Did he do anything? Or before dictating?"

Nurlan shook his head. "I came in only when he called yesterday evening and demanded I return to work. That's when he dictated the letter to you. He was in the lab all day and never left. He was talking to six of the Shepherds, but when I came, they left. After he finished the letter, he seemed very content. He was smiling, a real smile. I asked if he wanted anything else. He didn't speak for a while, just smiled and looked around and out the window at the lights on the playground. I wondered if he had taken some kind of narcotic. All he said was 'Wait. Wait' So I waited. Maybe half an hour. He stood there at the window a long time, then turned around, looking at me, but I had the sense that he didn't see me. I was just the furniture. Then he sat—almost fell—onto the couch and became angry. He began to repeat a name, Bulat. He said it over and over and over until he lay down and closed his eyes. I asked if I could go. He said, 'Go? Yes, go. Bulat's gone. The man Hank is gone. I am going. You go.' Then he said the names as if asking a question, 'Bulat? Henry? Arman?' He said that several times very quietly, then closed hie eyes and seemed to sleep. I left."

"Do you know anyone named Bulat?"

They both shook their heads. Then Michelle said softly, "Henry is the formal name for Hank. Hank was my husband's name."

"Maybe Bulat is one of the Shepherds?"

Asel shook her head and Nurlan said, "I've never heard him call any of them by that name."

In the moment of silence that followed, Michelle felt Akenov's wrist for a pulse. Faint, and slower than before. She had a lot to think about. Whether or not she had committed a crime, she knew she had to leave the scene, but she had to take Avalon and Aidana with her. Somehow she had to leave not as a criminal, but as a visiting American scientist with two daughters who were obviously Kazakhs, not Americans. But how?

Nurlan and Asel stood watching her, tacitly putting her in command.

She turned to Nurlan. "Who else knows about this?"

"I don't know. I don't know what happened between the time that I left and when I returned this morning."

"Good, then let's try to get out of here," Michelle said. "You two should go separately. You don't want to be associated with me, but I might need help later after I talk to Avalon. My cell phone is in my room, and I'll give you the number." She wrote it on a page from a memo pad on the desk, and wrote their numbers on a second page for herself. "Now please leave, as normally as you can make it look."

But when Asel opened the door, the little Shepherd stood looking at them, a large envelope under one arm, his hands in the pockets of his leather jacket.

"Sit down please," he said in slow and heavily accented English. It was a phrase that everyone in the Russian speaking world seemed to learn in the first week of English lessons. He added. "We have business. Nurlan translate."

The Shepherd spoke to him in staccato Kazakh, clearly used to giving orders.

Nurlan translated. "He says that he is following Dr. Akenov's wishes."

Michelle said, "I demand to call the US embassy."

Nurlan translated that, then listened to the little man's response.

"He says we can handle everything ourselves, or else they can call the police, and you will be charged with murder. He said you can call your embassy if you make the right decision. He believes Dr. Akenov has told you what your choices are."

Michelle knew she was not a good liar, so she would be honest but careful with her words. "Yes, he explained what he wanted my choices to be." She looked at Nurlan as she spoke. Just before he turned to the Shepherd and began translating Michelle, saw a very slight nod of understanding—not just of the English words but of her deliberate evasion of accepting terms.

When Nurlan had finished translating, the Shepherd, with a cruel little smile looked beyond Nurlan at Michelle and speaking in Kazakh, he said, "Dr. Murphy, do not play with me. The Good Shepherds are not ignorant followers or thugs. In fact, unlike you, we are what you might call enhanced members of the human species."

Michelle looked at him, her face showing surprise and skepticism, and he added, "Perhaps you knew that Dr. Akenov discovered and tested the gene KL-VS long before your Americans in California found it. It was his first attempt to transmit a specific gene to a human subject."

"He told me he had tested it on himself."

"Not just on himself."

"He gave it to all of the Shepherds?"

"Not all of us, just those who carry the genes of Genghis Khan. We are not so silly about equality as you Americans. We understand that we must have some people whose intelligence does not interfere with their willingness to do boring, sometimes risky work."

"Very smart," Michelle said with intended irony.

"Well then," the Shepherd said, "Let us sit down. I have my instructions. You are now the murderer of Dr. Akenov, and if you do not agree to the offer he made you, then you will become the murderer of everyone in this compound. That includes the three of you and the children here. How many would you like to kill?"

Asel said one word in Kazakh which Michelle guessed was the word "but" or something that meant the same. The Shepherd without looking at her said something harsh, and Asel did not speak again.

Michelle, thinking their conversation might be recorded, said, "I have killed no one and don't want to kill anyone."

Nurlan translated the terse reply. "Wrong. Sit!"

Michelle stood for a moment beside a chair facing Akenov's desk, then sat down. The Shepherd sat in Akenov's chair and carefully placed a document on the desk. "You will sign now."

"If we are going to do business," Michelle said, "tell me your name."

The man looked at her suspiciously, then said in his awkward English, "You call me Azbal."

"Mr. Azbal or Azbal what?" Michelle asked. She felt more in control demanding answers.

"Azbal Shepherd," he said in English.

Michelle stared at him for a moment, then leaned back in the chair away from the desk and the paper. "I won't sign," she declared and folded her arms in front of her.

"I think you will sign. I will give you time to think about this. Simple choice: you murder one man and you can go free and your daughter will be safe here. You refuse and you are charged with murder and you murder everyone in this compound."

"And how will I do that?" Michelle asked.

"As you can see by what happened to Dr. Akenov, you have brought with you, into his laboratory a fatal virus. That is already known fact. Perhaps you have brought other fatal biological agents with you."

Michelle began to protest, but he ignored her and continued.

"Dr. Akenov said you had a particular interest in a virus known as SV40. Everyone knows, of course, that the United States never abandoned biological weapons."

Michelle realized that he was telling her not what he believed, but what he and probably Akenov wanted the world to believe. She said, "I brought only two samples of a modified virus, not SV40. It was a virus Dr. Akenov didn't even know about."

"And with one of those you killed Dr. Akenov. The other viruses you brought . . ."

Michelle cut off the translation. "I didn't bring any other viruses. And I didn't kill Dr. Akenov. He took a great risk and killed himself."

Nurlan said, "I will finish what he said, then what you said. Azbal said, "The other viruses you brought include a specially modified form of SV40 virus." He paused an instant and translated into Kazakh Michelle's protest.

Azbal smiled. "Perhaps you forget. Perhaps you want to forget. You came here because you knew Dr. Akenov was an expert on this virus. You were afraid he was developing a way to transmit virus. Cause fatal results, but also a vaccine to protect. And that is true, so you killed him."

Michelle gathered her thoughts and tried to buy time and information. "What does this special form of SV40 do?"

"You know very well. After your attendance at the conference one of our young women who cleaned your room was killed by your virus—your first kill. Causes blood to clot everywhere in body. That is why Dr. Akenov developed a vaccine."

Michelle closed her eyes, feeling a wave of deep depression bury her senses. The last piece of the puzzle had fallen into place. Azbal had confirmed that Akenov had murdered Hank.

She forced herself to listen again to Azbal. He spoke as if he believed this story and Michelle began to think he did. Azbal smiled, savoring his winning game, thinking that Michelle had given up. "If I decide to release the virus tonight, you will see for yourself how well it works. But then you will also be dead along with all your victims. You will be, shall we say, a suicide biological bomber."

Michelle now understood the full scope of the fix, the frame up. She looked at Asel, then Nurlan. They understood, and they seemed curiously resigned.

Asel said very quietly, "You must sign everything. It's the way it is. This virus exists. They will use it. You have to do it for the children—for Avalon and for Marat. Please."

Nurlan had translated very quietly for Azbal.

Azbal picked up the papers and slid them back into the envelope. "All the documents will be waiting for you by . . . let's say, after supper. What do they say in your country, 'the condemned man ate a hearty meal?'"

Michelle looked at Nurlan. He gave a limp shrug, but Michelle noticed that he was quietly scrutinizing Azbal. She turned to Azbal. "Yes, I need some time to think. I have an idea that might be better for all of us. And I want to talk to my daughter."

Azbal said, "The choice is very clear, but so that you may be at peace with your choice, you have until 8 pm this evening. We are not without sentiment, and Kazakhs are known for their generosity. You and Avalon can have a final meal together. Then the children will have the play that they have planned for several weeks. It will be a very fitting end of your visit, however it ends. I shall leave one of our shepherds with you." He paused to let that irritate her, then added, "In case you need anything, just ask him."

Azbal opened the door, barked something in Kazakh and within seconds four black coated Shepherds turned a corner in the hallway, one pair behind the other and stopped outside the door. Michelle could not help staring at them. Two of the Shepherds were carrying military style automatic weapons, and the other two had pistols in holsters. Azbal spoke to them, and in a soft voice with a distinct tremble, Nurlan translated Azbal's orders for Michelle. He and Asel would remain in the lab. One of the Shepherds would take Michelle to her room where she would stay with Avalon and Aidana.

Michelle walked silently beside the Shepherd assigned to her. He was a young Kazakh with short brown hair, a gymnast's body, or a judo competitor's—loose, springy, and ready. He was also one of those with a pistol, although he seemed uncomfortable having it at his hip. Michelle was surprised by an almost motherly feeling for his innocence as he led her downstairs and through the playground to her apartment. At the door to Avalon's room he turned to her and indicated that she should enter. She was surprised when he said in English, "Your room. Go in please."

Michelle didn't go in. Instead she stared at the young Shepherd. "I know you."

The Shepherd shook his head and put a finger to his lips. "Shhhh."

Michelle mouthed his name, utterly astonished. *Bakhyt.*

The young man nodded, then placed his lips next to her ear and whispered, *Lipkovich. Explain later.* Michelle felt an enormous sense of relief. Evan was nearby. She had help. Maybe she would survive.

She nodded at Bakhyt and knocked on the door. Avalon opened it with a big smile, pulled her in and looked up at Bakhyt with a puzzled expression. Michelle hugged Avalon, whispering in her ear, "Yes, it's Bakhyt. We have help."

The young Shepherd stepped into the room and looked around quickly, then without saying anything pointed up at the ceiling, then at his ear. Avalon nodded, and Michelle realized that she had a new script to follow, to act as though Bakhyt was just another Shepherd sent to guard them.

Michelle asked, "Do you speak English?"

Bakhyt moved to the door, now also an actor in the play. "Little," he said. "You stay. I wait, keep safe." He went into the hall and pulled the door shut behind him.

Chapter 37

Michelle told Avalon the choices she confronted, and Avalon explained them to Aidana. They knew that Akenov was no longer listening, but perhaps the Shepherds were, so they needed to speak as though they were prisoners in their room, using body language to convey their real meaning. As she laid out the terms of each choice for Avalon, Michelle did not have to guard her words because there was no choice to make. The SV40 threat could be a bluff, but if she called the bluff and the Shepherds released the virus, weaponized by Akenov, she would have killed everyone including herself. And even if she signed the papers, she was now sure that the Shepherds would never feel secure with Asel and Nurlan alive. And for that matter, herself as well.

As if reading her mind, Avalon said, "Mom, we have another choice. There's something else we can do."

Michelle looked at her daughter, wondering how there could possibly be any alternatives, but she had long ago learned to expect surprises from Avalon. She waited.

"Dr. Akenov recruited many Shepherds, and he vaccinated them against the SV40 virus. Doesn't that make sense?"

"Go on."

"But I know he didn't vaccinate all of them."

"How do you know that?"

"Mom," Avalon said with a tender impatience. "You know—puzzle pieces, connecting dots. I can't say exactly how, but I understand him, and I have his plan in my head. He wants super Kazakhs, but he doesn't want weak links. He is obsessed with the idea that he is a direct descendent of Genghis Khan, carrying his genes, so he would have selected Shepherds with the genghisid markers to vaccinate. The rest are dispensable, so they would not be vaccinated."

Michelle began to understand. "But why do they all think they are vaccinated?"

"Easy. Just stick a needle in and say 'vaccinated'."

Michelle thought about what they were saying. Did it matter if the Shepherds were listening? She decided it did not. If they were, it would sow distrust. Divide and conquer. And if no one was paying attention, all the better.

Avalon knew her lines now, and was into her role. "We need to prove this to one of them. We can tell the guard outside that he is not immune."

"But suppose he is."

Avalon said, "He's not. He's too young. Akenov doesn't trust young people. We can test him the same way you tested Aidana and me."

Avalon quickly explained this to Aidana, then opened the door and saw the young Shepherd looking out the window at the end of the corridor. With a gesture and her best Kazakh she indicated she wanted to talk to him. "Very important," she said. When she took a step toward him he held up his hand, shaking his head, but walked over to stand in front of her.

Aidana had understood the plan and came to the door, looking over Avalon's shoulder. In Khazakh, she said they wanted to know if he was descended from Genghis Khan.

Bakhyt was puzzled by the question, but nodded. "All Shepherds inherit from Genghis."

"Did Dr. Akenov ever give you a vaccination?"

When he said he had, Avalon decided to gamble. "We know that only Shepherds who are descendants of Genghis Khan have been vaccinated against a deadly virus made by Dr. Akenov. To keep them safe if he ever needed to use it."

Bakhyt glanced at Michelle, then back to Aidana, shrugging. "He gave all Shepherds a special shot."

"That's what you were told, but it is not true," Aidana told him. "If you are not from the royal line of Genghis, you were not vaccinated. Not really."

The two girls looked at him and both nodded their heads, smiling. They were just old enough to be flirtatious.

"What proof?"

Avalon spoke up. "We can test whether your vaccination was real."

Although Lipkovich had convinced him that Akenov was dangerous and needed to be watched, Bakhyt still felt a certain loyalty to the Shepherds. He had never questioned the idea that Akenov instilled in them, that they were all special, working together toward a greater goal.

"What is the test?" he asked.

Avalon did not understand the entire exchange, but she could read his face. She said softly to Aidana, "Vaccination test first."

Avalon glanced at Michelle who gave the slightest nod. She went over to her carryon bag and took out the test kit, handing it to Avalon.

Aidana said to the Shepherd, "We can prove you are not vaccinated. If we show you, will you help us?"

Bakhyt hesitated. He swallowed, looked up and down the hall, then nodded quickly.

Avalon put the scraper and gel in Aidana's hand and Aidana told Bakhyt it was like the tuberculosis test all Kazakhs of his age had had. She pointed to the inside of her arm and to Avalon's. "See that red spot? That means Akenov vaccinated us to keep us safe. If you get a spot like that it means you were too. But if nothing happens, you were not. Want to try? It only takes a few minutes to show up."

He shrugged and told her to close the door, leaving Michelle alone in the room. He looked up and down the corridor again before quickly slipping out of one sleeve of his black leather jacket to expose his arm. He watched impassively while Aidana used the scraper to roughen his skin, then applied the gel and patch. "How long?" he asked.

"Ten minutes," Aidana said. "If it turns red under the patch, you have been vaccinated. If nothing happens, you are not immunized."

Exactly 10 min later he knocked softly on their door. Michelle opened it to see him standing there with his jacket off. He held out his left arm so that she could remove the patch. There was no inflammation. His skin was clear. Aidana stepped alongside Michelle and saw real confusion on the Shepherd's face.

Michelle now spoke in English. "You are one of many Shepherds who received a fake vaccination. Why?"

When Bakhyt did not answer, she continued, making her language as simple as possible. "No vaccination, then you die."

"How do you know this?"

"If I tell you, will you help us escape?"

Bakhyt hesitated, looking at her and at Aidana and Avalon. Michelle stepped back into the room, picked up the list that Nurlan had found and described for her. She gave it to Batyhk and said, "I can't read it, but Nurlan told me that this is Akenov's list of all the Shepherds. Do you see yourself in the Genghis list? It's on the first page."

Bakhyt examined the first page with only six names and shook his head. "The non-Genghis Shepherds are on the second page," Michelle said. Bakhyt looked intently at the labels on the second page, 20 names, then pointed to his.

"You understand now," Michelle said. "No Genghis, no vaccination. No vaccination, you die, the Genghis people live."

Finally Bakhyt nodded. "What can I do?"

Michelle closed her eyes and released her breath that she had been holding. Her motherly feeling came back stronger than before. "Cell phone," she asked softly.

He took a cell phone out of his pocket but shook his head, pointed to his ear, and with a finger made a big circle above his head to indicate others could listen in.

"Okay," Michelle said, thinking fast. She turned to Avalon. "Do you remember the number for anyone at the church?"

"Sure." Avalon stepped back into the room and dialed the number. When a woman answered she spoke rapidly in Spanish. "Por favor, urgente. Hay que hablar con Padre Carlos." She and Michelle had agreed that anyone listening was unlikely to understand Spanish. Or the Latin she remembered from high school. When father Carlos was on the line, Avalon spoke very quietly in Spanish, saying, "Por favor entiende que le llama usted el barco diciendo primero de Mayo tres veces. Tambien . . ." she then repeated very carefully Michelle's Latin phrase, "Da nobis intempore necessitates. Legatus." She pushed the off button. She turned to Michelle who was looking worried. "He said, 'Intelligo. Fiat voluntas tua.' He got it. If any of the Shepherds are listening in, it will take them a while to get it, if they do." The Spanish had said, *Pease understand that the boat is calling you saying, May Day, May Day, May Day.* The Latin had said, *We need immediate help. Embassy.*

Bakhyt looked in with a worried expression and held his hand out for the cell phone. He looked at Aidana and said, "Now they will come. Close the door." Within minutes they heard quick footsteps in the corridor and two men talking rapidly to Bakhyt, then a slapping sound. Aidana listened to the conversation with her ear to the door.

With her hands, a bit of English, a bit of Kazakh and pantomime, she explained that "Bakhyt told them we had picked his pocket of his cell phone and he got it back as soon as he heard Avalon talking. They took his cell phone away, called him an idiot and slapped his face."

A moment later they heard a soft knock on their door. Aidana opened it slightly and listened, then closed the door. "He wants to know what to do next."

"The other names on the non-Genghis list," Avalon said. "There are more of them than the others, almost four times as many."

"That's right," Michelle said. "Only 15–20 % of Central Asians have Genghisid markers. "So how do we make our Bakhyt the leader of his own army?"

Avalon smiled. "The play's the thing, wherein we'll wake the conscience of the khans."

Chapter 38

Within minutes of talking to Avalon, Father Carlos had called the American embassy in Astana. The charge d'affairs had listened politely, asked him to spell his name, repeat his phone number, and had assured the priest that "the appropriate people will be informed."

When Father Carlos had asked, "You do believe me, don't you?" the official had replied, "I have no reason to doubt that you believe what you say." Father Carlos had heard the same words several times in his life about the existence of God. After smothering his own anger, his parting words were, "Thank you. I have no reason to doubt you will do what you think is right." And even as he turned off his own phone and slipped it into his robes, he said to himself, *And I will do what I think is right.*

At the embassy the U.S. Charge d'Affaires immediately left his office and walked through the windowless corridors to the office of Lt. Colonel Bill Southern, the military attaché. Southern was immediately on the phone with Evan Lipkovich. He was surprised to hear Lipkovich say that he had arrived in Almaty that morning and had already driven past Good Shepherd to get some idea of what they were up against. "The place is locked down and guarded by Akenov's Shepherds. No sign of Akenov himself. We're told this is not the usual guard-on-duty crew. Something different is going on, and I think we might have a hostage situation. By the way, the weapons the bad guys have are pretty nasty, not just guns. This Akenov has a long history of research on biological warfare agents."

"I can have some people down there in three hours," Jakhontov said.

"Thanks, but not yet," Lipkovich replied. "The compound is in a village outside the city where everybody knows everyone else. There's only one road in and I'm sure they are watching it."

Lipkovich paused for a moment, then said, "My one local asset hasn't been heard from. I assume he's inside without any way to communicate. We can't get close to the compound without letting them know they are being watched. I'm worried what might happen, because Akenov is a highly intelligent sociopath."

"What if I talk to the President's people I know—military. The word will go up to the top fast."

"Then what?" Lipkovich asked.

"Say what you want about the president here, but he listens to his generals. And he doesn't take kindly to having someone else butt in on his power."

"Do it." Lipkovich said.

Avalon and Aidana called Bakhyt to the door and had a quiet but intense conversation with him. It was now 5 in the afternoon, and Azbal had granted the students one hour of rehearsal time for their play. Before the girls left for the rehearsal with Bakhyt, they told Michelle what role she would have and what her cue would be later in the evening.

Like schools everywhere, the dining hall was designed to be turned into an auditorium with a stage at one end. The rehearsal was carefully guarded. Bakhyt joined the Shepherds when they arrived, and Avalon quickly counted 26 of them standing near the walls, obvious in their black uniforms, six carrying automatic weapons, the rest with holstered side arms. The girls had only one small change to make in the script. Avalon had no trouble passing a note to Marat and saying a few words while they stood on the side watching the others rehearse their songs and recite verses from Makataev and Abai and other Kazakh writers.

When the other children began to file into the dining at the usual time for supper, they were at first curious why so many men in black were there, but soon ignored them as they ate and chattered excitedly about the play. Then supper was over and the students carried their dishes to the collection window, then returned to their seats and turned their chairs to face the stage. The Shepherds brought even the kitchen staff to watch. With the players assembled in a small room to one side of the stage, Azbal walked front and center to the microphone and stood with legs slightly apart, hands behind his back, a smug smile spread across his face as he waited for the audience to stop talking. Two Shepherds brought in extra chairs for a short front row. When they had finished, Azbal tapped the microphone to be sure it was on. The loud thumps quieted the last of the chatter.

"We welcome everyone here for a special night and a special celebration. Now I would invite our special guests to come in."

As the guests, escorted by a Shepherd on each side, came in, Azbal introduced them.

"Father Carlos of the Holy Roman Catholic Church, has decided to join us tonight." Michelle only knew Father Carlos from what Avalon had told her, and was surprised to see the tiny priest in his brown robe come forward. The little Franciscan smiled at her, then nodded when a Shepherd motioned for him to take a seat in the first row of chairs.

Next Azbal announced, "Mr. Zautbek Kunaev, one of our own citizens but also an employee of America's Central Intelligence Agency. We thought he would be more comfortable viewing from inside than watching our walls from outside."

Kunaev, as short as Father Carlos and dressed in shabby clothes, was hunched over and kept his eyes down as he was brought forward by the Shepherds. When he was seated, another Shepherd escorted Nurlan and Asel to the front row.

"And finally," Azbal announced with the confident smile of a man arranging an elaborate joke, "Please welcome the mother of one of our students and our special honored guest, the American scientist, Professor Michelle Murphy."

He led the students in applause as two Shepherds led Michelle to the front row where they motioned her into a seat next to Nurlan. Azbal held up his hands to stop the applause, then made a sweeping gesture toward the side of the stage and shouted, "Let the play begin!" He walked off quickly and took a seat to one side where he could easily watch both the play and the front row visitors.

Michelle found herself watching something very familiar—a nativity pageant. Instead of Bethlehem, a stable and a manger, this savior had just been born on the floor of a shepherd's yurt. Instead of a star in the east, a bright light shined down through the central vent opening of the roof and bathed a swaddled doll with a brown face.

A student in cardboard armor, with a wooden spear came on stage and told the story of the famous "Golden Warrior" prince born to local Sythian tribes over 2000 years ago. Archeologists uncovered the 18 year old in a burial mound 45 miles from Almaty in 1969. There he lay in his gold plated armor, with his gold dagger and sword at his side and a tall pointed hat on his head. Around him lay the skeletons of his horses and clothing stitched with gold. The student concluded the story saying, "He who touches the Golden Warrior will feel his power. And the great scientist Dr. Arman Akenov has promised us that his power will be again the power of all Kazakhs."

The students clapped loudly. The Shepherds standing guard around the room also clapped. Azbal smiled from his seat and looked triumphantly at Michelle. She ignored him.

The play continued with students in costume reciting the deeds of the Golden Warrior's successors. The theme of each was the greatness of Kazakh people and their heroes. The last character was Marat who represented the future of the Kazakh people. Wearing a beige camel hair Kazakh robe and a tall conical felt hat, Marat walked quickly to the center of the stage. In one hand he held a rolled up scroll. He announced in a loud voice, "I am the future of Kazakhstan." He looked around the room slowly before he spoke again. "I am all who have descended from the line of the great Khans." He spread his arms wide, and at that gesture Avalon and Aidana swept towards him from opposite sides of the stage, wearing flowing Kazakh dresses and bright red rounded hats with a tuft of eagle feathers rising from the center. As they came to his side he dropped his arms and said, "These are my wives and from us shall come the new citizens and rulers of Kazakhstan." A ripple of applause spread across the audience. Michelle looked at Asel, surprised, but Asel smiled and whispered, "It's just a play."

"For now," Michelle said. Then, with some hesitation, she added, "but who knows? If we survive."

Avalon stepped forward and said in Kazakh, "I must say important words from my other language." Looking down at Michelle and Father Carlos she recited,

There is a tide in the affairs of men.
Which, taken at the flood, leads on to fortune;
Omitted, all the voyage of their life
Is bound in shallows and in miseries.
On such a full sea are we now afloat,
And we must take the current when it serves,
Or lose our ventures.

Michelle smiled a proud motherly smile to anyone looking at her. But as yet she didn't know what she would be proud of. She lip synced the reply that Shakespeare's Cassius gave to Brutus, "Then, with your will, go on." Avalon bowed to her and to the audience, then stepped back behind Marat.

Marat unrolled the scroll, and looking over the top of it he addressed the audience. "I declare that you should know that I, Marat, am the true son and heir of Dr. Arman Akenov. As some of you already know, my father died this morning." The audience let out a collective gasp. The Shepherds standing along the wall already knew. They showed no trace of emotion but surveyed the audience critically for any possible disturbance. Marat continued. "It was my father's wish and his life's work to create a race of Kazakhs superior to all other races and fit to rule by virtue of intelligence."

Michelle looked at Azbal sitting to one side of the stage. He sat upright with his arms folded in front of him a small but confident smile on his face, but a watchfulness and a certain tension showed in his posture.

Marat continued. "Tonight, our play adds a new act in the history of the Kazakh people and the world. History and evolution have chosen some to rule and others to be ruled and yet others to be eliminated for the good of everyone." He began to unroll his scroll and hold it before him to be read. "My father wanted me to thank all of his Shepherds. I will read one set of names. Aidana will read another set."

Azbal rose to his feet and stared without knowing what to expect.

Marat's short list of six names began with Azbal and when he was done, he bowed and amid applause, he handed the list to Aidana. Azbal turned and surveyed his men spaced around the walls like black posts. Aidana read a second list of names longer than the first. She too bowed to applause and handed the list back to Marat who rolled up the scroll and stepped forward to speak. "In a few minutes, the leader of my father's Good Shepherds who are here with us tonight plans to release the agent of racial evolution, a deadly virus." He spoke very fast as he saw Azbal striding toward the stage. "The names read by Aidana are those Shepherds who are not descendants of the Khans, and they have false vaccinations. They will die. With all of you."

Azbal ripped the scroll from his hands and ordered him and the girls off the stage. They retreated to one side where one of the larger and more brutish Shepherds stopped them. "ATTENTION!" Azbal shouted above the commotion that had broken out in the audience. "Stay in your seats!" He looked with menace at the adults in the first row. They looked around the room at the Shepherds along the walls and entrances. They understood and remained sitting.

Azbal went on. "These young people have made a very poor joke. They have turned a serious drama, an homage to a great man and a great idea—they have turned it into a farce." He shifted into his most sympathetic voice, his own act. "Of course, I hope none of you were frightened by this nonsense. Now let me tell you the good news. I have several important announcements that will interest you all." He paused, shifted his position, tried to rise taller than he was, and straightened his jacket. "First, I will tell you that the true enemy sits before me in the first row." He drew an envelope out of his tunic and held it up. "I have already received the confession of Professor Michelle Murphy that she has brought into our school a deadly virus that she intended to use." He pointed to Michelle. "No one is vaccinated against the biological agent she brought here and with which she murdered Dr. Arman Akenov."

All eyes turned to Michelle, but she did not turn around.

Akenov continued. "The play that was presented tonight was corrupted and made a farce. Although Dr. Akenov cannot be here, he has prepared what we might now call the final act." Azbal paused and took note of the attentive and quiet audience.

Avalon and Aidana studied him intently from their position off stage. Michelle caught Avalon's eye and she thought Avalon nodded to her very slightly.

Azbal turned his head to where the two girls waited with their Shepherd minder and ordered Avalon go sit next to her mother so that Nurlan could translate for both of them.

Michelle was surprised to see that as Avalon approached, her daughter seemed to have a slight smile on her face. Michelle said nothing as Avalon took her seat beside Nurlan.

Azbal said, "If all of you will be patient for just a few more minutes, we have an important new safety drill. It is made necessary by the kind of threat that Michelle Murphy brought into our country." He nodded to a Shepherd standing just off stage and the man brought him one of the school's fire extinguishers and a gas mask. Azbal cradled the extinguisher in his arms and said, "Let us pretend that this fire extinguisher is one of Professor Murphy's vaporizers, full of deadly viruses of the kind she intended to use. If you were not vaccinated against such a virus, you would need a protective mask. Watch while several of our Shepherds and I demonstrate how you would use a mask, although I am quite sure you will never need it. I will do mine first."

With that, Azbal slipped the mask over his face. Three of the Shepherds closest to the stage and two behind the audience took from their jackets flexible gas masks. As they pulled them on, Azbal pointed the extinguisher up in the air, pulled the trigger and a spray of white mist shot out over the audience.

Avalon stood up and shouted in Kazakh, "Жер сілкіну! Жер сілкіну!!"

Immediately all of the well-drilled students dropped to the floor in their earthquake drill routine and took refuge under their seats. Asel quickly did the same, followed by Nurlan and Avalon who pulled Michelle down with them. And at that instant Bakhyt, realizing that his lack of a gas mask was the final proof that he was dispensable, pulled from his jacket an automatic pistol. He raised it quickly and fired at Azbal who fell sideways, dropping the fire extinguisher. The children screamed, startled by the unexpected explosion, but their screams were drowned out by the clattering noise of automatic weapons as confused Shepherds fired at Bakhyt and each other. Father Carlos saw that Azbal, bleeding from a wound in his shoulder, was still groping for the fire extinguisher. He scrambled in a crouch toward the stage with bullets zinging overhead and short tattoos of firing coming from one direction, then another. Father Carlos rolled onto the stage, snatched the extinguisher, and hugged it to his body as he rolled away from Azbal, who was no longer moving.

"Get out!" Bakhyt yelled to the remaining Shepherds, all without gas masks. "The white smoke is poison!" Doors banged and the room became quiet except for the sounds of children crying, some still holding their ears.

On the floor Michelle said, "Don't breathe. We need to get out of here. Fast." She tried to pull Avalon and Aidana to their feet.

Avalon said, "Mom, it's safe. It was just a fire extinguisher with CO2, like we have at home."

Michelle, frantic, was still pulling her up, trying to get her to run. "Akenov loaded it with SV40!"

"It was CO2, mom. Believe me! I figured it out long ago. Fire extinguishers have twins too. I told Marat to switch the extinguishers."

Michelle, dazed, with ears still ringing from the gunfire, just stood there, looking down uncomprehendingly at Avalon and Aidana. Then she burst into uncontrolled gasping sobs and fell to her knees, where her two daughters hugged her tightly. The other children were coming out from under tables, many of them still crying or simply standing in stunned silence looking at the stage and several dead Shepherds with gas masks still on. From somewhere among the children Mariam began singing a soft rolling melody. At that moment the doors burst open and Kazakh commandos burst into the room, weapons ready. Aidana and Nurlan simultaneously shouted in Kazakh and Russian, "Don't shoot! It's over!"

The soldiers paused, surveying the room carefully, then at a command from their officer pointed their guns toward the ceiling and relaxed. On the stage, father Carlos struggled to sit up but fell back. More troops were now pouring in, and in the confusion Avalon slipped away to the stage where she knelt by Father Carlos, the body of Azbal behind her. A dark spot in the priest's brown tunic was growing larger. Before she could speak, he said to her in Spanish between teeth clenched in pain, "Quick—take confession."

Avalon was confused. "You want to confess, you want me to confess?"

She thought she heard a sound like a laugh before he said, "No, from this man—your mother's confession."

Avalon understood immediately. Hoping no one would see her, she crawled away from father Carlos to Azbal and felt inside his jacket, first on one side, then on the other and found an envelope slippery with blood. She crumbled it into a wet ball and turned back to father Carlos. "I have it."

"Very good," he said faintly and closed his eyes. Troops were now everywhere and two of them knelt over Father Carlos. They called for a stretcher and soon two men in camouflage were carrying him away.

One of the officers was speaking through an electronic megaphone to the students, directing them into the courtyard. Michelle, Avalon, Aidana, Nurlan, Asel and the CIA man had gathered in a small group on stage, where an officer and six troops surrounded them. The officer sounded courteous when he spoke to Michelle in Kazakh, and Avalon translated.

"Mom, he wants us to follow him outside." The officer led them through the playground, soldiers trailing behind, then through the gate and outside into the quiet street. A full moon had risen, and Michelle saw a tall figure beckoning to her. When she got closer, she was astonished to see that it was Evan Lipkovich. Michelle did not know whether to punch him or hug him.

So she hugged him. "Evan! I can't believe it! Thank you thank you thank you!" Then she pulled away. "Can we leave? I never want to see this place again."

"There's a plane waiting for us at the airport."

"What about the others?"

"They'll be safe now with the Kazakh security people."

"Bakhyt? How is he involved in all this?"

"I guessed as soon as you told me about Bakhyt that he was sent by Akenov. Remember that we had a little talk with him when Rhonda and I visited? I told him what we knew about Akenov and gave him a choice. I could arrest him as a terrorist spy, or I would let him stay on as one of the Shepherds if he agreed to help us get you out. He's a smart kid, and he chose to help. But speaking of getting out, didn't you just tell me you wanted to leave?"

Michelle laughed and nodded, looking over to Avalon who was standing nearby listening. Lipkovich put an arm around Michelle and another around Avalon and began urging them toward a large black Ford Explorer. "Let's go. We were lucky to get permission for the plane to land, but the sooner we take off the better."

Avalon suddenly twisted out of his embrace, and when he reached for her, she said something in Kazakh, then in English with an accent, "I not Avalon. I Aidana. They are Avalon." She pointed behind her to her twin sister.

Lipkovich was confused. He turned to Michelle. "What the hell!"

Taking the hint from Avalon, Michelle knew exactly what to do. "Both of you girls, get in, now." Michelle immediately got into the back seat of the car, leaving the door open and beckoning. By the time Lipkovich realized there were two Avalons, the girls had locked arms, their body language saying both of us, or neither of us. Lipkovich glanced over to Michelle, who was smiling at him, eyebrows raised. Shaking his head, he guided the twins in next to Michelle and shut the door. Then he got into the driver's seat and looked back at Michelle, who was still smiling happily.

Evan said, "This is crazy."

"You ain't seen nothin' yet," Michelle replied.

Epilogue

The Year 2075

When Michelle opened her eyes, she realized she had been awakened by the recorded sound of a flute softly playing a lovely melody she recognized immediately, translated by Aidana from the FOXP5 base sequence. She glanced up at the wall panel opposite her bed that usually displayed a three dimensional Oregon scene, often a live image of a seascape or a waterfall on Mount Hood or even an icescape on Jupiter's moon Europa. This morning she saw gold letters shimmering against a deep blue background:

Now you're one hundred.
Oh, for the world you made
Your name will never fade,
Who ever wondered?

Sleepy as she was, Michelle had to smile at the game Avalon still played with poetry. The verse was a rift on Tennyson's Charge of the Light Brigade,

> *When can their glory fade?*
> *O the wild charge they made!*
> *All the world wondered.*
> *Honor the charge they made,*
> *Honor the Light Brigade,*
> *Noble six hundred.*

The Charge of the Light Brigade at three o'clock in the morning? She heard a soft knock, then Avalon and Aidana slipped in and stood just inside the doorway, beaming at her.

Avalon said, "Sorry to wake you, Mom, but we just had a call."

Michelle yawned and rubbed her eyes. "I'm sure we're going to have a lot of them, but really, one hundred is just another birthday, just another day. Would you mind answering the phone? There aren't many people I want to talk to at three in the morning."

"Mom," Aidana continued, "you have no idea how many people are going to call today."

"Who was it?" Michelle asked.

Aidana and Avalon looked at each other with broad grins. Avalon said, "No one you know."

Michelle raised both palms in the air in surrender. "But you are going to tell me, right?"

Together they repeated the verse in gold letters:

> *Now you're one hundred.*
> *Oh, for the world you made*
> *Your name will never fade,*

Who ever wondered?
Nobel 200.

Aidana nodded to Avalon, who said, "Mom, you may want to call these people back. They had a Swedish accent."

"Why? I don't know any Swedes."

Her two lovely middle-aged daughters giggled like teenagers. Avalon said, "I told them you were asleep, so they asked me to pass along a bit of news. Are you awake now?"

Michelle was getting exasperated. "Yes I'm awake! What's this news that's so important."

The twins had rehearsed. Avalon said, "Mom, listen closely: "Who ever wondered, Nobel 200." They watched Michelle's face as she heard the difference between noble and Nobel. In unison Avalon and Aidan said, "The three of us will share this year's Nobel Prize in Physiology and Medicine. It's the 200th Nobel Prize in physiology and medicine." Aidana added, and you are the only woman besides Marie Curie ever to receive this prize twice in a lifetime.

The news articles that followed the announcement over the next few days often began with the personal story of the University of Oregon researcher who, at age 51, was awarded the Nobel Prize for the discovery and dissemination of the FOXP5 gene. Then a routine genomic analysis performed during her annual physical found that Michelle had a mutation in her BRCA gene revealing an ominous 80 percent risk of developing breast cancer. Many women who received this news opted for a prophylactic mastectomy, but her two daughters, both biomolecular engineers at the University of Washington School of Medicine, persuaded her to join them in searching for a cure based on her pioneering research with TransTek Inc. in Eugene. The three of them produced a viral vector that homed in on malignant cells and switched them into apoptosis, the signaling pathway initiating cell death. Michelle began to treat herself with the vector, and underwent the usual annual mammograms and ultrasound diagnostic procedures. There was no sign of tumors, but her daughters began to notice a surprising side effect. As a woman in her 50s, Michelle was resigned to the usual wrinkles that had begun to appear on the skin around her mouth and on her forehead. But three years after she began the viral therapy Avalon and Aidana corralled her and made her stand with them in front of the full length mirror in her bedroom.

"Mom, look at yourself," Avalon said. "You're getting younger!"

Michelle nodded. "I've noticed it too, but I didn't dare believe it."

Aidana laughed. "We think we know what's happening. And it's amazing."

As sometimes happens with serendipitous discoveries, the twins realized that the viral vector Michelle fabricated also caused senescent cells to enter apoptosis. As they died and were reabsorbed by the normal turnover processes of metabolism, they were replaced by the stem cells present in most tissues of the body. Michelle became the first woman to have her skin truly rejuvenated as the newly born dermal cells began to replace damaged collagen and elastin fibers, so that age-related wrinkling slowly disappeared.

When Avalon and Aidana published their discovery in *Science*, with Michelle listed as corresponding author, reporters had called it "the New Fountain of Youth" even though the authors warned that the aging process had only been slowed, not reversed. Stem cells would inevitably be depleted. But even with that caveat, when the tabloids and blogs and Internet news services wrote about the two daughters who had rejuvenated their mother, they had no need to "hype" the stories. The simple truth was sensational enough, and now the news cycle would begin again with the announcement of the Nobel Prize.

Three months later, in early December, Michelle and her daughters were waiting with the other laureates in the VIP lounge of Stockholm's Concert Hall. Some old friends were sitting in the audience. Professor Julie "The Cooler" Flanagan was there, still freckled and red haired and attentive as she had been many years ago when she sat next to Avalon as Michelle lectured about Ma, the first human with the *FOXP2* mutant gene. Evan Lipkovich sat next to her, his full head of snow white hair glowing in the spot lights that illuminated the first row of seats. He had risen to become deputy director of the CIA and instituted a variety of new and effective policies to deal with biological warfare and bioterrorism. His former colleague, Rhonda Grable, had died of ovarian cancer years ago, shortly after Michelle and the twins escaped from Kazakhstan with Evan's help. Sergey Smagulov was also in the front row of invited guests. After Akenov died and Good Shepherd closed, he helped Asel get the children back to their parents, then moved from Almaty to London where he worked for many years translating Russian poetry and literature. Next to him was Marat Satpayev, who finished high school in Almaty, then went on to undergraduate and graduate studies at the University of Oregon. He completed his PhD in Biology with Michelle as his research advisor. She told him about her early interest in astrobiology, so with a new PhD in hand Marat became a staff scientist at the Jet Propulsion Laboratory in Pasadena where he flourished. He became the principal investigator of a proposal to NASA for a sample return mission to Jupiter's ice covered moon Europa. Against all odds, the mission was successful, and five years ago Marat became the first scientist to nurture the single celled life forms retrieved from the probe. And next to him, completing the arc of old friends sat Botagaz, who had emerged from her orphanage childhood to become Kazakhstan's distinguished Minister of Science.

Michelle requested that two seats should remain vacant between Lipkovich and Sergey to honor the memory of Jodie and Don Koskin. Don had continued to work with Kazakh scientists to unearth more remains of ancient Denisovans, then used bioinformatic analysis to prove that they had come to Central Asia from Africa in a previously unknown wave of nomadic people. The year after their famous Denisovan paper appeared in *Nature*, Don and Jodie perished in 2025 while kayaking along the Oregon coast. The San Juan Plate slipped, creating a 9.3 Richter scale earthquake that sent a fifty foot high wall of water crashing onto the coast and roaring up the bays and rivers. Their bodies were never recovered, but their twisted carbon fiber kayak was later found in the great Sea Lion Cave north of Florence. Another 30,000 Oregonians also lost their lives in the tsunami, most of them disappearing with the backwash into the Pacific Ocean.

Michelle had requested that Avalon and Aidana should be first to give their acceptance speeches. Avalon began with a clear summary of all the changes science had brought to humanity in the last half century. For the first time in human history, she pointed out, the appearance of a new human species had not resulted in violent confrontations, as it had between Homo sapiens and the neanderthals. Instead, the new humans with FOXP5 had honored the old humans with *FOXP2*, as children honor their parents. Furthermore, the new humans, despite scattered terrorist attacks by religious fanatics, not only honored their parents but cared for them as they aged, providing the gift of longevity and the healing of common diseases.

Aidana was next to speak, focusing on the methods they had used to fabricate the viral vector that inactivated her mother's mutated BRCA1 gene. She included a personal note, describing her lonely childhood in a Kazakh family with no siblings, and the extraordinary circumstances that brought her together with her twin sister. She spoke of how the FOXP5 version that Julie discovered in Avalon's baby tooth had been made available to the entire world. "Everyone here today can now have that gene if they wish. How can Avalon and I be lonely when we are here with so many people who are our genetic relatives." She ended by reminding the audience how mythology can shape our culture. "Over two thousand years ago," she said, "the Greeks recognized that god-like powers in human form can give rise to catastrophic conflict, such as the war between the Titans and the Olympians. More recently the English poet Milton wrote an epic poem describing Satan's rebellion against God. We can continue to learn from those stories."

Aidana looked over at Avalon with a warm smile. "My sister and I were united by a man, a brilliant scientist, who desired god-like powers. But he wanted to use this power to divide the human race into two species. His life ended in tragedy. He was a victim of hubris, a word invented by the Greeks to warn about the consequences of pride. But our two species are now united. We are beginning a new chapter in the history of humanity, but we should never forget we are the same species that acquired speech only 200,000 years ago."

Avalon then rose and stood next to Aidan, taking her hand. She ended their part of the program by quoting Hamlet:

"What a piece of work is a man! How noble in reason, how infinite in faculty! In form and moving how express and admirable! In action how like an angel, in apprehension how like a god! The beauty of the world. The paragon of animals. And yet, to me, what is this quintessence of dust?"

Avalon paused for a moment, a rhetorical trick she had learned from Michelle, then went on. "Once upon a time we were nothing but stardust. Now we have all the power that myths and scriptures once assigned to gods. We have created new life forms and we have conquered most diseases, even the one called aging. We have engineered the very climate of the earth. We can guide the orbits of dangerous near Earth asteroids into new paths. We have colonized another planet and found life on a moon of Jupiter." She looked fondly at Marat who smiled back and felt his eyes fill with tears. "But we have not yet found intelligent life anywhere else. For now, at least, we are the gods. But even though the universe belongs to us, we must be careful, we must be humble. It's a beautiful universe that we must care about, and care for the Earth, our tiny speck of it. We must be wary of hubris."

Michelle stood up and stepped between Avalon and Aidana, a small white figure between the two tall and lean brown women. She spoke for just five minutes, the traditional, obligatory, and heartfelt words of a new Nobelist. After thanking the Queen of Sweden and the Nobel committee, she looked up first at Avalon, then at Aidana but let silence stand long enough for the audience to begin to whisper. Then she said "Do you know. . ." and paused, getting the audience to pay attention, "Do you know if these are my daughters? After all, I gave them none of my genes. But I am here today because they gave me their love, and their genes. Perhaps I am their daughter." After a very brief second, the audience burst into applause. Michelle spoke louder, making herself heard over the clapping. "Science made that possible," Michelle said, "but that act was not an act of science." She paused and took a deep breath. She went on, forcing the words through her emotion, presenting each as if it were a stone being laid in a foundation. "Their act was. . ." Tears were streaming down her face. She put her arms around Avalon and Aidana and pulled them close. She coughed and she said, "Well, I don't have to tell you."

That was all she said. And what she did not say became the subject of long discussions around the world.

Part II

The Science Behind the Fiction

Scientific Appendices

Kazakhs and Kazakhstan, a Brief Introduction

Neither archeology nor history have revealed when the Kazakh people came together as a distinct tribe or nation. Limited DNA analyses show that the Kazakhs share the DNA of many people who came to Asia as long as 40,000 years ago. This includes DNA common among American Indians. Their language has its roots largely in the Turkic language group shared by many people of Central Asia, including Mongols. In the 1400s the Kazakhs emerged as a distinct group led by princes of the Mongol "White Horde." By the mid 1500s three hordes (or zhuses) came under the rule of a single Khan. Outsiders began to write of Kazakhs in the 1600s as Russians began to explore and claim the vast expanse of desert, steppe, and forest north of the Himalayas and geographically part of Siberia. As the Russians built small settlements and projected military power, Kazakhs staged rebellions but later relied on Russian help in the mid 1800s to protect their lands from Uighurs, Uzbeks, Kirghiz and other Central Asians. By the time the communist Bolsheviks came to power the lands of the Kazakhs encompassed an area about four times the size of Texas, stretching from the Caspian Sea in the west to China and Mongolia in the east and lying below that part of Russia known as Siberia.

During the early years of Lenin and then of Joseph Stalin in the Soviet Union many of the Central Asian republics (Uzbekistan, Kyrgyzstan, Tadjikistan, and Turkmenistan) suffered from politically created starvation, especially the Kazakhs. The slow influx of Russians that had begun in the mid 1800s increased along with other ethnic groups that Stalin exiled en masse, especially Soviet Germans. In 1897 the population had been 74 % Kazakh, but by 1939 it was 30 % Kazakh, mostly living in the rural areas as herders. The Kazakh population had grown to 40 % in 1991 when the Soviet Union collapsed and Kazakhstan declared independence. By this time, many who had moved to urban areas no longer spoke much or any Kazakh.

© Springer International Publishing Switzerland 2016
W. Kaufman, D. Deamer, *The Hunt for FOXP5*, Science and Fiction,
DOI 10.1007/978-3-319-28961-8_2

Kazakh ethnic identity, however, had remained strong. In 1986 when the Soviet prime minister Gorbachev removed the Kazakh governor of the republic, Kazakhs staged a brief rebellion that the Communist Party put down. However, the new Russian governor was replaced with a Kazakh communist, Nursultan Nazarbayev who held onto power and became president of the new country. Nazarbayev, the son of illiterate sheep herders, became a metal worker and then a loyal communist, rising in the Soviet system. A new constitution declared it a crime to insult or demean the office of president, and Nazarbayev has continued to rule, having won several controversial elections. In 1994 he moved the country's capital from the southern city of Almaty (also called Alma Ata, meaning ancestor of apples) to a small town in the windswept steppes of central Kazakhstan, an area then dominated by ethnic Russians. There, using abundant income from its oil reserves (among the ten largest in the world), he built a new capital, Astana.

President Nazarbayev has said that his model for national development was Singapore, where an authoritarian leader gradually guided the country to more democratic rule and developed its economy and rapidly created a modern standard of living. Nazarbayev declared that Kazakhstan would join the ranks of the modern developed countries by the year 2030. As part of that program the President instituted the Bolashak Scholarship, which sent hundreds of Kazakh students to America and other countries for higher education.

While Astana has become a successful capital city, Almaty, population 1.5 million, continues to be the "commercial capital" of the country. It is bordered on the south by the 15,000 ft snow covered range of the Tien Shan (Celestial Mountains). On the north the greenery nourished by melt waters and rains gives way quickly to desert, then farther north to the flat Siberian steppe, and in the northeast, the beautiful Altai mountains.

For many reasons large numbers of non-Kazakhs have left the country. Speaking Kazakh became a requirement for higher office in the 1990s, and all legal documents are now required to be in both Russian and Kazakh. The leading positions in the country's largest firms are now held almost entirely by ethnic Kazakhs who make up some 64 % of the population of almost 17 million, with Russian ethnics claiming 24 %. Although the word Kazakh indicates ethnicity and the word Kazakhstani signifies any citizen or institution of the country, more and more institutions bear the adjective "Kazakh", a trend that irritates some of the minorities in the country. Journalists have generally followed this practice. It is, in fact, akin to calling institutions in Europe or North America "Caucasian".

President Nazarbayev supported legislation in 2010 that exempted him from the law that says a president cannot serve more than two terms. He turned 75 in 2015 but did not announce any political plans to either run again or to retire.

Ma and Mitochondrial Eve (Chapter 2)

A "clock" telling time ticks in the genomes of all living organisms. The ticking is not mechanical, but instead it marks time by the rate at which mutations occur in the DNA. By counting and comparing the number of mutations that have occurred in the genomes of different species, or different individuals within a species we can estimate the amount of evolutionary time that has passed.

In the human genome, this clock marks time in two places: the nuclear DNA which is inherited from both parents, and the mitochondrial DNA which is inherited only from the mother, because sperm do not contribute mitochondria when they fertilize an egg. In the human mitochondrial genome, one mutation occurs every 40 generations (800 years), which is about ten times faster than the mutation rate in nuclear DNA.

Although estimates of the rate at which mutation clocks tick vary widely, it is at least possible to assume that it falls within a certain plausible range. Using this number we can deduce how much time has passed during the evolution of an organism from its most recent common ancestor. Allan Wilson and his graduate students Rebecca Cann and Mark Stoneking at UC Berkeley were first to measure mutations in human mitochondrial DNA and were able to conclude that all human beings carried a mitochondrial genome descended from one woman who lived in Africa around 140,000 years ago. She became known in the popular literature as Mitochondrial Eve. This does not mean that she was alone in the human population at that time, just that her DNA happened to exist in an unbroken lineage to this day. It is reasonable to assume that she had the human version of *FOXP2*, and this is why Julie and Don were so excited when they found it in the human tooth Julie brought back from Ethiopia.

FOXP2 (Chapter 7)

Language and intelligence are interwoven. Individuals who do not do well on intelligence tests typically use simpler language to express themselves, and higher scores are associated with more nuanced and complex language. The first indication that there might be a genetically controlled aspect of language was reported in 1990 by Myrna Gopnik at McGill University who studied a British family (referred to as KE for anonymity) half of whom suffered from an inherited disorder that inhibited their ability to produce speech with coherent grammar. She suggested that this might have a genetic explanation, and in

1998 Simon Fisher and his research associates reported that the family had a defective gene on chromosome 7 that they named SPCH1. When the sequence was analyzed, it was found to correspond to a family of proteins called forkhead box transcription factors, so the protein was named *FOXP2*, and the gene then became *FOXP2* (by convention in italics).

The idea that a gene and its protein product might be associated with language skills stimulated great excitement, and over the next decade the *FOXP2* gene was found to be universally distributed among mammals, with other versions also present in birds and related to their singing ability. Svante Pääbo and his research group compared the human protein with that of chimpanzees, and reported in 2002 that the human version had two amino acids that were different, one of which was unique to humans. In a later paper, the human version was also found in the Neanderthal genome. Pääbo warned that this does not mean that Neanderthals were capable of complex language, because many other genes are also involved.

Given that simple mutations in *FOXP2* can have such a powerful effect on speech, it seems plausible that another mutation in a FOX gene could directly affect intellectual ability in humans, perhaps by increasing the number of synaptic connections per neuron in the cerebral cortex. In fact, research by Richard Huganir and Gek-Ming Sia at the Johns Hopkins University found that a gene called SRPX2 could increase synaptic connections in mice. The premise of *FOXP5* is that Avalon had a mutation in one of her FOX genes that increased her intelligence beyond that of ordinary humans, to an extent similar to that of the mutation in a transcription factor that allowed the first humans to become a separate species of primates 200,000 years ago in Africa.

Viruses and Cancer (Chapter 3)

More than 100 years ago a young physician named Peyton Rous discovered a virus that could produce sarcoma cancer in chickens and proved it could be spread from one bird to another. It took years before other scientists accepted his work, but in 1966 he finally received the Nobel Prize. The Rous sarcoma virus (RSV) was the first to be shown to cause cancer.

In 1960, another virus was discovered in the tissue cultures of monkey cells that were being used to produce polio vaccines. It was named simian vacuolating virus (SV40) because of the large empty vacuoles that formed in infected cells. This was particularly disturbing because Bernice Eddy later showed that sarcoma cancers occurred in hamsters if they were injected with cells containing SV40.

Yet another cancer causing virus is the human papilloma virus (HPV) that was first discovered by Peyton Rous in warts but is now known to cause cervical cancer. Young women are urged to be treated with HPV vaccine as a prophylactic against cancer.

Given that viruses can produce cancer, Evan Lipkovich suspected that Arman Akenov might be testing viruses as a way to carry out assassinations without anyone suspecting that the death was caused on purpose. That would explain why he ordered the SV40 virus sample.

Blood Clotting (Chapter 4)

When the SV40 viral vector did not work as a cancer-inducing agent, Arman Akenov devised another approach, also based on SV40. He used it on Hank Murphy to take revenge for Hank's revelation that the artifacts being sold by Akenov to support his early research at Good Shepherd were fakes. Azbal also threatened to use the SV40 based weapon in his showdown with Michelle in the last chapter.

How could a viral vector be designed to induce uncontrolled blood clotting, a highly complex process that is normally under tight control? The final step in blood clotting occurs when an enzyme called thrombin acts on fibrinogen and turns it into fibrin, which in turn forms long fibers that entangle blood cells and form the clot. However, the process begins when an injury to tissue exposes platelets to a connective tissue protein called collagen. The platelets release activating factors such as ADP and serotonin into the localized site of injury and these in turn initiate a cascade of events that ultimately turns prothrombin into thrombin and leads to clot formation.

Akenov found that he could target an SV40 preparation to bind to surface receptors on platelet membranes after it was inhaled as a fine powder. The virus activated the platelets and initiated clot formation, leading to a fatal outcome as multiple thromboses blocked blood flow throughout the body. Hank became the first victim a few hours after he opened the package sent by Akenov through his Swiss antiquities dealer.

The Power of Genetic Mutations (Chapter 7)

When one sees the base sequence of a gene for the first time, it looks like some sort of encrypted code, nothing but A's, T's, G's and C's in apparently random order. Here is one example taken from the human genome that affects the life of every human being depending on whether or not it is expressed as a protein during embryonic development:

```
  1 gttgaggggg tgttgagggc ggagaaatgc aagtttcatt acaaaagtta acgtaacaaa
 61 gaatctggta gaagtgagtt ttggatagta aaataagttt cgaactctgg cacctttcaa
121 ttttgtcgca ctctccttgt ttttgacaat gcaatcatat gcttctgcta tgttaagcgt
181 attcaacagc gatgattaca gtccagctgt gcaagagaat attcccgctc tccggagaag
241 ctcttccttc ctttgcactg aaagctgtaa ctctaagtat cagtgtgaaa cgggagaaaa
301 cagtaaaggc aacgtccagg atagagtgaa gcgacccatg aacgcattca tcgtgtggtc
361 tcgcgatcag aggcgcaaga tggctctaga gaatcccaga atgcgaaact cagagatcag
421 caagcagctg ggataccagt ggaaaatgct tactgaagcc gaaaaatggc cattcttcca
481 ggaggcacag aaattacagg ccatgcacag agagaaatac ccgaattata agtatcgacc
541 tcgtcggaag gcgaagatgc tgccgaagaa ttgcagtttg cttcccgcag atcccgcttc
601 ggtactctgc agcgaagtgc aactggacaa caggttgtac agggatgact gtacgaaagc
661 cacacactca agaatggagc accagctagg ccacttaccg cccatcaacg cagccagctc
721 accgcagcaa cgggaccgct acagccactg gacaaagctg taggacaatc gggtaacatt
781 ggctacaaag acctacctag atgctccttt ttacgataac ttacagccct cactttctta
841 tgtttagttt caatattgtt ttcttttctc tggctaataa aggccttatt catttca
```

This sequence codes for the testis determining factor, abbreviated TDF. The remarkable fact is that out of 3 billion base pairs in the human genome, the presence of absence of that short sequence, just 897 bases long, determines whether an infant is a male or a female. Human cells have two strands of DNA arranged in the now famous double helix. The DNA is present in 23 pairs of chromosomes, but these are separated during meiosis to produce haploid sex cells called sperm and ova, which then contain 23 unpaired chromosomes.

During meiosis chromosomes get stirred up in a process called crossover, and there is a chance that bits of genetic sequences turn up on the wrong chromosome. For instance, suppose that the TDF sequence happens to cross over from the Y to the X chromosome during sperm development. When that sperm unites with an ovum, it brings into the egg its X chromosome with TDF attached. The resulting embryo will produce a male infant, but one that is genetically female by virtue of having not XY but two X chromosomes.

The other possibility is that certain mutations in the TDF sequence can inactivate it. Although an embryo might have the XY pair, once the instructions for testes are turned off, the bearer grows up to be a normal female. This is known as the Swyer syndrome, and the first instance was reported in England in 1991. It is extremely rare, and only one in 10 million females are estimated to be genetic males.

These examples show how a miniscule change caused by crossover or mutation can affect human life. Another powerful effect is the mutation in the FOXP2 gene that led to the origin of the human race, which Michelle described in her lecture to Don Koskin's class in Chapter 5.

Intelligence and Cortical Complexity (Michelle's Lecture in Astana, Chapter 13)

There have been dozens of attempts to define and explain intelligence and consciousness, but there is as yet no consensus about an underlying mechanism. The only certainty is that consciousness emerges from physical and chemical processes in cerebral neurons, but how it emerges remains one of the great questions of biology. In her lecture, Michelle was bold enough to propose a quantitative measurement of intelligence that could be applied to mammals ranging from mice to humans. She argued that consciousness and intelligence are best understood in terms of an evolutionary process that began when animal life developed differentiated cells called neurons. The evolution of the nervous system is a central theme of FOXP5 and will be discussed in detail here.

During the Cambrian radiation between 580 and 500 million years ago, the fossil record shows the appearance of small animals now called *Bilateria*. It is reasonable to assume that these organisms had nervous systems, but it is uncertain whether they represented the nervous function that later evolved into the nervous systems of today's animals. No doubt it would have been a major selective advantage for the predators and prey of that era to be able to sense their environment and respond with appropriate behavior. The chief characteristic of this level of nervous function is a *reflexive sensory-motor response* to sensory input. This basic function is preserved in higher organisms as well, for instance in the spinal reflex response to a painful stimulus, as Michelle pointed out in her lecture.

The next step came with the ever increasing complexity of animal nervous systems as life evolved into larger aquatic organisms like fish and then into terrestrial animal life 400 million years ago. When we compare the behavior of fish, reptiles, birds and mammals, it is clear from observations that the vertebrate animals are increasingly aware of their environment. Instead of being entirely reflexive, their responses to sensory input can be modulated within certain limitations. Their modulated responses apparently reflect a short term memory measured in seconds, which cannot support truly intelligent behavior.

The behavior that characterizes *self-awareness* arose in the nervous systems of primates and other large-brained animals such as elephants and dolphins. A self-aware organism recognizes itself in a mirror, and *Homo neanderthalensis* 400,000 years ago would likely have had no difficulty passing this test. Self-awareness evolved into modern consciousness 200,000 years ago with the appearance of *Homo sapiens* in Africa. If a child from that era could be transported forward in time to the present, it would presumably be indistinguishable from other children in its ability to develop language and adapt to contemporary culture.

The most striking property of a conscious human being is not just self-awareness, but to varying degrees human brains can indefinitely maintain an internal model of sensory input and manipulate the model in order to consider possible future outcomes. Short term memory is therefore not measured in seconds, but instead can be maintained throughout a problem-solving interval. The word *intelligence* defines a semi-quantitative measure of the ability of the conscious nervous system to perform such tasks.

Michelle presented a set of postulates that can be used to clarify the discussion of human consciousness. The postulates, taken together, also suggest experimental and observational tests of hypotheses related to consciousness. The first postulate is that consciousness will ultimately be understood in terms of ordinary chemical and physical laws. This conclusion is supported by the fact that the conscious state is strongly affected by chemical and physical conditions imposed on the brain. For instance, consciousness is abolished by lowering the temperature of the brain by $10°$. When the brain is warmed, consciousness returns. A similar effect is produced by general anesthetics which diffuse from the lungs into the blood, then partition into cell membranes of the brain and interact with protein channels such as GABA and glutamate receptors. Anesthetics are specific examples of a large number of chemicals that interact with receptors in the cell membranes of cerebral neurons and thereby produce effects ranging from the mild stimulation of nicotine and caffeine to deep anesthesia. If small amounts of such chemicals interacting with neurons can reversibly affect consciousness, it seems inescapable that the mechanisms underlying consciousness most likely involve biochemical and physical processes occurring at the level of cortical neurons and their interactions with one another.

The second postulate is that consciousness is related to the evolution of anatomical complexity in the nervous system. Michelle noted that the reason consciousness seems so mysterious at present is that we have not advanced far enough in our knowledge of complex interactions within the brain's neurons. This is analogous to the evolution of computer engineering over the past 70 years. Imagine that somehow a functioning laptop computer could be transported back in time to Los Alamos in 1943, where some of the worlds most brilliant physicists had gathered in wartime to design and test the first nuclear weapon. Although the ENIAC computer with 17,000 vacuum tubes was being developed at that time in Pennsylvania, most of the scientists' calculations at Los Alamos were being done with pencil and paper. They would have been astonished by the laptop's color screen, the fact that an entire movie could be stored, the WiFi capacity, and internet access. No matter how brilliant, their collective genius would be baffled by this seeming miracle. Michelle noted that we are like those scientists when today we attempt to understand how the phenomenon of consciousness emerges from nervous function in the brain.

The second postulate also suggests that consciousness can emerge only when a certain level of anatomical complexity has evolved in the brain that is directly related to the number of neurons, the number of synaptic connections between neurons, and the anatomical organization of the brain. Again by analogy to the evolution of computers, a certain number of components and interacting connections are required to perform increasingly complex tasks. Consider the evolution of the integrated circuit. The first IC was developed by Kirby and Noyce in the 1950s, and incorporated only a few semiconductor-based transistors. In the late 1960s the number of transistors in an IC had increased to 100 s, then to thousands in the mid-1970s. The number increased again to the 100,000 range in the 1980s, to millions in 1990s, and most recently billions. Each advance represents a threshold relating the number of transistors and their connections to the complexity of computational function.

Michelle compared this history to the evolution of the nervous system. The earliest animals were well served by a nervous system having perhaps a few hundred neurons. The different cell types in *C. elegans* have been counted: there are precisely 302 neurons and a total of 7000 synaptic connections. In contrast, the human cerebral cortex is estimated to have 10–20 billion neurons and a total of $\sim 10^{15}$ synapses. If in fact consciousness and intelligence are related to complexity of nervous systems, it should be possible to establish a quantitative measure of the complexity, then compare it with our observation of animal behavior.

This brings us to Michelle's third postulate, that consciousness, intelligence, self-awareness and awareness are graded, and have a threshold that is related to the complexity of nervous systems. Michelle proposed a quantitative formula that gives a rough estimate of the complexity of nervous systems. Only two variables are required: the number of units in a nervous system, and the number of connections (interactions) each unit has with other units in the system. The formula is simple: C (complexity) $= \log(N) * \log(Z)$ where N is the number of units and Z is the average number of synaptic inputs to a single neuron.

The idea that complexity arises from interconnecting systems is not a new concept. W. Grey Walter suggested much the same thing in his book *The Living Brain* published in 1953. Michelle restricted her list to a set of mammalian species for which estimates of cortical cell number and synaptic junctions per cell are available.

The number of neurons (N) increases markedly within the nervous systems of animals ranging from mice to humans. The number of synapses per neuron (Z) also varies significantly. Z is difficult to estimate, but has been measured for cortical neurons in the human, rat and mouse brains. The numbers vary by a factor of 2 within the six cell layers of the neocortex, and again by a factor of 2 when the human, rat and mouse brain are compared in terms of average number of synapses per neuron for all six layers. Each human cortical neuron has approximately 30,000 synapses per cell, each mouse neuron 20,000 synapses per cell and each rat neuron 17,000 synapses per cell. For purposes of her calculation, Michelle assumed that Z = 30,000 for the brains of primates, dolphins elephants and monkeys, and Z = 20,000 for the brains of all other animals in the list. For the order of magnitude calculations reported here, these rough estimates of Z are sufficient. Table 1 shows the list of mammals ranked according to the number of cortical neurons and log (N). The last two columns show log(Z) and the value of C calculated as log(N)*log(Z). Brain weight is also given for comparison.

Michelle noted that the order of mammals in this list matches what most people would conclude from their personal experience and observations. All six animals with normalized complexity values of 43 and above are self-aware according to the mirror test, while the animals with complexity values below

Table 1 Mammals ranked by number of cortical neurons (N) and synaptic connections (Z) per cell

Animal	Brain weight (grams)	Cortical neurons (millions)	Log(N)	Log(Z)	C
Human	1350	11,500	10.1	4.5	45.5
Elephant	4200	11,000	10.0	4.5	45
Chimpanzee	380	6200	9.8	4.5	44.1
Dolphin	1350	5800	9.7	4.5	43.6
Gorilla	480	4300	9.6	4.5	43.2
Horse	510	1200	9.1	4.3	39.1
Dog	64	610	8.8	4.3	37.8
Rhesus	88	480	8.7	4.3	37.4
Cat	25	300	7.6	4.3	32.7
Opossum	7.6	27	7.4	4.3	31.8
Rat	2	15	7.2	4.3	31
Mouse	0.3	4	6.6	4.3	23.4

40 do not exhibit this behavior. The discontinuity between C = 40 and 43 appears to reflect a threshold related to self-awareness.

Although mammals with normalized complexity values between 43 and 45 are self-aware and are perhaps conscious in a limited capacity, they do not exhibit what we recognize as human intelligence. It seems that a normalized complexity value of 45.5 is required for human consciousness and intelligence, that is, ~11.5 billion neurons, each on average with 30,000 connections to other neurons, and an EQ of 7.6. Only the human brain has achieved this threshold.

What kind of mutation would produce the next significant increment in human intelligence? From the argument above, it seems likely that a mutation like *FOXP5* could lead to small increases in the number of cortical neurons and the number of synaptic connections per cell. For instance, to increase the C value by the same amount as the difference between elephants (45.0) and humans (45.5), in other words, to 46.0, an increase in the number of cortical neurons from 11.5 billion to 12 billion, and an increase of synapses per cell from 30,000 to 37,000 would be sufficient. If a mutation like *FOXP5* slightly increased the number of cells and synapses it could lead to a significant increase in intelligence such as Avalon displayed.

It is interesting to compare Michelle's results with another quantitative way to compare mammalian brains. One such is called the encephalization quotient (EQ) and is illustrated in Fig. 1. Note that the ratio of brain weight to body weight of humans far exceeds that of other animals, which simply means that the selective processes related to human evolution favored increased brain mass and presumably greater intellectual power. However, the other points on the graph seem less satisfactory. For instance, the mole and the rhinoceros have

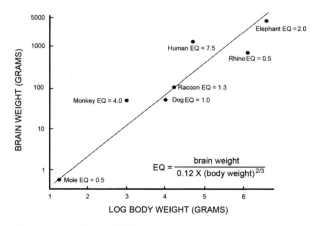

Fig. 1 Encephalization quotient (EQ) for large and small animals

the same EQ, but it is likely that according to Michelle's formulation the rhino will be more in line with other large mammals.

Viral Vectors (Chapter 30)

Paul Berg, who shared the Nobel Prize in 1980 for his research on recombinant DNA methods, realized in the 1970s that viruses could be genetically manipulated and used to introduce novel genes into mammalian cells. In his biography, Berg wrote: "Soon after I returned to Stanford, I conceived of using SV40 as a means for introducing new genes into mammalian cells much in the way that bacteriophage transduce cellular DNA among infected cells. My colleagues and I succeeded in developing a general way to join two DNAs together in vitro; in this case, a set of three genes responsible for metabolizing galactose in the bacterium *E. coli* was inserted into the SV40 DNA genome. That work led to the emergence of the recombinant DNA technology thereby providing a major tool for analyzing mammalian gene structure and function and formed the basis for me receiving the 1980 Nobel Prize in Chemistry."

Viruses reproduce by delivering their genetic instructions and a few enzymes into living cells. The viral genes, which can either be DNA or RNA, then take over the protein synthesis machinery of the cell and produce viral proteins and nucleic acids that assemble into hundreds of new virus particles. When the cell dies, these are released and go on to infect other cells.

It was obvious to Paul Berg that viruses could be a powerful tool for gene therapy, and thousands of papers now describe how certain viruses can be used as viral vectors. The most common are adenoviruses, adeno-associated viruses, and retroviruses called lentiviruses. All viruses work by binding to receptors on the cell surfaces of specific tissues such as the upper respiratory tract (colds and influenza) liver cells (hepatitis A, B and C viruses) motor neurons in the spinal cord (poliomyelitis) or neurons in the brain (rabies). Michelle Murphy and Arman Akenov found ways to alter this property so that a virus would bind to and deliver genes to other kinds of cells. Michelle considered the possibility of delivering *FOXP5* to egg cells in the ovaries of young women. If the egg was fertilized and produced a pregnancy, the infant would presumably grow into an adult with Avalon's advanced intelligence. Because the viral vector containing the gene was a modified cold virus, the *FOXP5* gene would spread to the entire human population within a single generation to produce the next human species. In her confrontation with Akenov, Michelle had to decide whether to give him the viral vector, or perhaps release it herself in order to stymie his plans.

CRISPR/Cas System (Chapter 30)

The CRISPR/Cas9 technique used by Michelle to insert *FOXP5* into a viral vector is a highly efficient new way to edit genetic material. CRISPR (pronounced crisper) is an abbreviation for Clustered Regularly Interspaced Short Palindromic Repeats. In 1987, these were discovered in the genomes of *E. coli* bacteria by Yoshizimi Ishino, but only years later did it become apparent that they are related to a kind of immune system that protects bacteria from infection by bacteriophage viruses. In between the CRISPR sequences are spacer sequences, short fragments of bacteriphage DNA that have been incorporated into bacterial genomes over millions of years. The CRISPR system works in conjunction with Crisper Associated genes (abbreviated *cas*), which code for enzymes called nucleases that cut DNA. When a bacteriophage injects its DNA into a bacterial cell, the foreign DNA is recognized and targeted by the RNAs coded by the spacer sequences that exactly match the phage DNA. The nuclease then proceeds to cut the viral DNA to deactivate it. In 2005 three research groups successfully used the CRISPR concept to develop synthetic guide RNA sequences that could target specific sites in the DNA of virtually any organism. When the guide RNA was added in the presence of the *cas* nuclease along with a DNA fragment to be added to the genome, the nuclease cut the DNA at the desired site and a second repair enzyme would add the DNA fragment.

Made in the USA
San Bernardino, CA
13 March 2020